Contents

Chapter One: Why become a beekeeper?/How to set up and get started in beekeeping 4

Chapter Two: Getting to know your bees and management of your hives of bees in the spring 19

Chapter Three: Management of your hive 22

Chapter Four: The Honey Harvest 41

Chapter Five: The June Gap 47

Chapter Six: Travelling with Bees 50

Chapter Seven: Preparing for winter 53

Chapter Eight: Pests and Diseases 55

Chapter Nine: Dealing with Beeswax 60

Glossary 62
Index 65
Frequently asked questions 66

Chapter One:
Why become a beekeeper?
How to set up and get started in beekeeping.

There are all kinds of reasons why people take up beekeeping. Probably the most common reason is the satisfaction of harvesting your own honey and wax and the enjoyment of eating or using your own produce. Many people become beekeepers because they need the excellent pollination which honeybees provide for fruit blossom and most vegetables with the resulting increase in yields. Bumble bees are not only picturesque but excellent pollinators too, but bumble bees are relatively few in numbers compared with honeybees, and of course they give us no honey or wax. Some people become beekeepers because the craft is already in the family. They have the advantage of available experience and mentoring. Others want to have bees because it is fashionable, or their friends are beekeepers, and beekeeping associations, of which there are many, have interesting meetings and provide opportunities to meet others. A hive or two can be an attractive addition to your garden or farm, and people are fascinated by the ingeniousness of bees which seem to demonstrate a universal intelligence in all life. The bees can even be pets and many people love to watch them as they disappear into the blue sky and go we know not where, or watch them as they appear, and drop down to the hive entrance laden with nectar or pollen.

A few people become beekeepers in order to make a living at it. One famous commercial beekeeper, during a local beekeepers' association visit to his apiary said, 'Nowadays you need to have a thousand hives to make a living at beekeeping'. He employed several people, had dedicated transport and a modern factory to deal with the harvest. Since the arrival of the varroa mite in the 1990s in Scotland, it is likely that he requires even more hives to make a living because the disease threat to honeybees is now so great. A few years ago, before the onslaught of the varroa mite, the author calculated that his beekeeping work paid a similar rate per hour to what a gardener was often paid. The time devoted to the weekly inspection of the hive in summer is only a small part of the time devoted to the craft. Equipment must be prepared, the harvest has to be dealt with, and when in the finished state, the produce must be labelled and marketed or given away to friends and relations. The very efficient beekeeper who mentored the author had around twenty five hives, and he used to say at the end of the summer season, 'I can't complain, because I've got my ton of honey, and it will pay for my winter in Portugal.' It must be said that few amateur beekeepers can hope to get that sort of crop from their bees. My mentor worked practically every day on his bees, and he also moved his hives to take advantage of forage at other sites. He also pointed out that more hives did not always mean more honey: 'What's the sense in having ten hives with a poor yield because you fail to look after the bees properly when you can have an equal amount of honey with five hives which are well managed?'

Will you have time for beekeeping?

Beekeeping is a fairly time consuming activity. One can reckon on spending at least one hour per week for each hive during May, June and usually part of July. The attention of the beekeeper is required on particular days, and unless the weather is very bad, usually no other day will do instead. That time must be set aside, because the beekeeper is bound by the bees' timetable which cannot easily be altered except by the bees. Failure to follow the bees' timetable and take action will usually result in the loss of bees in a swarm, and little or no honey for that year unless the beekeeper is fortunate enough to find the swarm, take it, and hive it.

Careful timekeeping with the bees' calendar is the essence of successful harvests. A notebook should be kept, with details of the particular stage the bees have reached with regard to swarm preparations and other important developments such as when a new queen can be expected to be mated. A person who is away from home for days on end in the middle of summer is unlikely to be a successful beekeeper unless he or she has a fully competent deputy. The hour is not usually important, but the date is often critical.

If you become a beekeeper.

Clearly, it is advisable to keep only as many hives as can be managed well with the time and other resources available. It must be pointed out that bees can be a considerable nuisance to other people if they are not looked after properly. The most obvious nuisance is when a hive throws a swarm or even several swarms and the swarms alight in your neighbour's garden or part of the neighbour's house. A good beekeeper is able to control his bees' swarming behaviour to a large extent, even if no beekeeper has one hundred percent control over his bees. One of the things you will soon learn to do is to catch a swarm (see page 15).

How to start beekeeping.

The first requirement is a hive to provide the bees with a home. The hive will require frames, and beeswax foundation inserted into the frames upon which the bees will build their combs consisting of thousands of hexagonal cells in which larvae are raised, and pollen and honey are stored. Since the nineteenth century the movable frame hive has become the universal method of keeping bees efficiently, and the barbarous practice of killing the bees for their honey in the autumn is, thankfully, now no longer practised. Nowadays, using our knowledge of the regular pattern in which the bees use their cells, we can encourage the bees to build up large stores of honey above the frames and cells in which they rear larvae and store pollen. A queen excluder is placed over the frames for rearing larvae, the 'brood frames', in the 'brood chamber', fitted out with 'brood' or 'deep' beeswax foundation which is of a larger size than the foundation used for the honey 'super'. Above the queen excluder you generally place one or more 'shallows' or shallow crates, filled with shallow frames fitted with 'shallow' beeswax foundation. The shallow crate is usually called a 'super'. Without the queen excluder, to separate brood rearing from honey storage, the queen bee would, of course, lay eggs in the shallow combs, and honey, larvae or 'brood' and pollen would become less easily separated. To get started, the novice beekeeper will require a hive equipped with a 'deep' or brood box for the brood, a queen excluder and a 'shallow' box or 'super' for honey. To prevent the bees building comb from the roof, the shallow box will require to be covered with a 'crown board' or 'cover board'. The bees' house is completed with a floor and a roof. When you have got the hive completed, fitted out with frames and foundation, you will require a suitable site. The best aspect and situation for a hive is generally with the entrance facing south east in order to catch the morning sun. The hive should be a reasonable distance from hedges and trees in front in order that the hive should get a little winter sun, which will warm the hive slightly and make it easier for the bees to move about and find their honey stores within the brood chamber. It is also advisable to situate the hive well away from large trees for safety reasons. You do not want to have your hive smashed by a falling branch. Away from trees is also best, because you do not want your hive to receive drips from branches which will shorten the life of the hive and could lead to damp accumulating in the hive. Hedges and trees to the north, west and east of the hive, at a reasonable distance, will help to provide shelter from strong winds. Similarly, walls and buildings at a reasonable distance from the hive can provide a little shelter, but careful observation should be made in windy conditions because solid structures can give rise to powerful back-drafts and eddies.

Other equipment you will need to get started.

A hive needs to be kept off the ground in order to keep the floor as dry as possible. Large baulks of square timber can be used, a number of bricks could also suffice, and sometimes old plastic crates are employed although they are not very attractive. Special hive stands with legs are even better, but must be stable. A strong hive tool is essential for separating the hive parts. The author finds the broad hive tools are stronger, narrow ones can bend too easily. A good smoker of medium size is probably more useful than the large ones which could become rather heavy and clumsy; the author prefers copper smokers because galvanized steel smokers can become rather rusty in time. A veil is usually essential to prevent stings to the face and eyes. Gauntlets or gloves are useful if the bees become angry. The author generally prefers to work without gloves or gauntlets because handling bees and frames is then much less clumsy, and the bees get used to the hands of the beekeeper without distress to the bees or the beekeeper. Cuffs or even elastic bands are useful to prevent bees crawling up the arms. Two pairs of socks tucked into trousers will generally prevent stings to the ankles which can be troublesome occasionally. In order to expose only one or two frames at a time when going into the brood chamber, and thus keep the bees warm and better tempered, it is useful to make up manipulating cloths. Four dowels about a half inch in diameter, the same length as the hive along the frame tops are needed. Two pieces of canvas of the same dimensions as the hive across the top are glued to the dowels effectively making a pair of roller blinds with the canvas rolled around the dowels. You can even make do with only one canvas roller, and have an old sheet rolled up of suitable size for the other manipulating cloth. The sheet roll is soft, which facilitates very gently placing the flat of the frame on the sheet when catching the queen, and if you are slow and gentle the bees will not get squashed.

What are the costs of beekeeping?

Unless you can buy your hives and equipment second-hand, getting started in beekeeping is a fairly expensive business. At the time of writing, a new hive complete, but without frames and foundation can leave little change out of three hundred pounds. Add frames, foundation and queen excluder at another twenty five pounds. Add a smoker, a hive tool, a veil and gloves and depending on the type chosen one can say goodbye to another hundred pounds at least! Then, there is more expensive equipment to deal with the harvest. How you are going to deal with the harvest will have an important effect on how much you need to spend on equipment. Probably the cheapest option is to produce cut-comb honey, but yields will be significantly lower than if you are producing extracted honey. That is because the bees have to build new comb to store the honey each year, whereas with extracted honey production the combs, usually wired for strength, can be used over and over again, which results in a heavier harvest. Each pound of wax requires the bees to consume several pounds of honey! One can spend around three hundred pounds for an extractor, which separates the honey from the comb by centrifugal force. Sometimes, however, honey cannot be extracted because the honey is too viscous or worse, granulated or crystallised, and in the latter case the honey is solid in the combs. Such a situation can only be remedied by putting the frames into a special heater for a period in order to melt the honey again, or alternatively, by cutting the honey out of the frames. You will need buckets, a honey strainer, and a container with a honey tap for filling jars.

If you live in an area where oil seed rape is grown, you could dispense with the extractor and buy a water heater into which you put the buckets of honey comb and then you melt the honey out of the combs which are smashed up with a clean stick. Oil seed rape honey is very viscous, and granulates or crystallises very rapidly, which means that it is often impractical to extract it as mentioned above.

If you are going for cut comb production, use of new foundation every year is an extra expense, and normal thick foundation is unsuitable for cut comb because it is too chewy. If you are intending to produce cut-comb honey simply allow the bees to build natural comb on thin foundation. You can even dispense with full sheets of thin foundation and just use 'starter strips' which are strips of thin foundation about an inch wide and may be 'glued' along the top bar of the shallow frame with a little melted beeswax, or attached in the normal way. The starter strip helps to avoid the tendency of the bees to build the combs all 'higgledy-piggledy'. Cut comb honey is generally popular and commands a higher price because it is less plentiful, and yields of honey are relatively reduced with this method of production. Cut comb honey for market requires scales to weigh the exact amount of honey in the plastic food box. Honey already granulated and solid in the combs is not usually regarded as suitable for sale as cut comb honey.

About sixty years ago much honey was sold in 'sections'. These are generally square whitewood frames, with both faces open and section wax foundation in the middle onto which the bees build their combs and store the honey. Sections are relatively expensive to buy and unlike extracted shallow combs cannot be used repeatedly. Section honey is therefore more expensive to produce, and it is said that the bees are much more reluctant to work on sections than they are to fill ordinary super combs. Sections are likely to have the effect of interrupting the cluster of bees which suggests that heavier yields can be expected where the cluster is less interrupted in a full width super frame. See page 43, on dealing with the honey harvest for more information.

What type of hive should I get?

Bees can be kept successfully in many types of hive. Bees are very adaptable and can make a home in the most unlikely places. For efficient beekeeping it is important to keep to one type of hive, and not mix types because it is sometimes necessary to put boxes one on top of another, for example, when uniting colonies of bees, and it is sometimes necessary to take frames from one hive and put them into another hive. Be warned, however: never unite bees without taking the precaution of putting a sheet of newspaper between the boxes. When the bees have chewed through the paper gradually, they will be able to make friends. Otherwise the beekeeper will probably be horrified to see the bees fighting each other to death. An acquaintance of the author's put two colonies together without knowledge of that precaution and lost most of the bees in appalling carnage!

At the time of writing there is a fad for the cheaper Top Bar hive. This hive is adopted by many people in the view that it will be preferred by the bees, and is more natural. These hives take top bars only, so that there are no frames to carry the combs which are simply attached to the top bars, there being no side bars or bottom bars of wood to support the combs. The bees build natural comb downwards from the top bar, and the best sort of top-bar hive is wider at the top than at the bottom, so that the bees build their combs with the natural catenary similar to the way a necklace hangs or an anchor chain descends to the anchor on the sea bed. The idea is that the bees will not then attach the combs to the sides of the box. That may well be, but it is very likely that often they will build their combs in a higgledy-piggledy fashion across the direction of the top bars, although that may be minimised if 'starter strips' of foundation are attached to the top bars. The major drawback of the top-bar hive is that the combs are very fragile without the support of side and bottom bars, and if the combs can be taken out of the hive at all, they should never be held horizontally or the weight of the honey or larvae will snap the comb off the top bar with consequences which could be a small disaster and is not likely to favoured by the bees or appreciated by the beekeeper! The top-bar hive is a retrogressive type of hive because it is the type of arrangement which existed before Langstroth's invention of the movable frame hive in about 1850. In the modern hive, the frames are surrounded by the built-in bee space of the design, which mostly prevents the bees attaching the frame sides to the hive walls.

With the top-bar hive, regular inspection of your combs is likely to be much more difficult on account of the fragility of the combs, and the fact that they are more likely to be built in an irregular way unless great care is taken to insert a board at regular time intervals, leaving only one comb space after the previous straight comb, thus forcing the bees to build their new comb parallel to the previous comb. That, however, is not entirely 'natural'. If combs are not regular and parallel to each other, it can become nearly impossible to remove the combs without damage to them. In any type of hive, regular inspection of the combs is essential for disease prevention, sometimes for disease treatment, also, for finding the queen, for placing the queen on her comb into another hive during swarm control measures, for checking that the hive has sufficient honey stores for its needs and that the queen is laying and the hive is doing all right. The list could go on… The only advantages of the top-bar hive are its simplicity and cheapness. In practically every other respect this type of hive seems likely to create difficulties for the beekeeper. Consider, for example, what is likely to happen if it is necessary to shake the bees off a comb for the purposes of creating another hive, or for disease control. In such operations, the comb will almost certainly break off its top bar. Consider also that in the top-bar hive, the bees are only able to store their honey in the same box as the larvae and pollen and not in an easily separable box above the brood nest over a queen excluder as in the modern hive. It is the observation of virtually all beekeepers that the bees naturally store their honey in a canopy, above and down the sides of the brood nest. As far as the author is aware, the top-bar hive gives little or no opportunity for the bees to store honey above them in any quantity, so they are forced to store their honey at the back of the top bar hive, thus making no advantage of the bees' natural food 'canopy' pattern. Sensible beekeepers, who respect their bees, never take the honey which is in the brood nest. The view is that the honey in the brood box is 'theirs'. Generally this means that nearly all the brood frames have a canopy of honey at the top, and the two frames immediately inside the walls of the modern hive are usually full of honey, 'their' honey as stated. Above the brood nest is the rent the bees pay the beekeeper, your honey harvest! The author strongly recommends that beekeepers adopt the modern movable frame hive and should not make un-necessary difficulties for themselves by reverting to an earlier type of hive. If the beekeeper wants his or her bees to build natural comb, the place to do that is in the honey super. The author generally has all natural comb in the supers because of the prevalence of oil-seed rape in the lowlands, (see dealing with the honey harvest, chapter 4) and the human preference for natural comb in producing cut-comb heather honey when the bees go on their 'busman's holiday' to the highlands. At the time of writing, the current vogue for the top-bar hive appears to be based on the idea that the modern hive, being less 'natural', is unkind to the bees and is a factor in the current problems facing the bee population. The current fad may also go with the idea that bees are stressed by being interfered with by the beekeeper. In the author's experience, over thirty six years of beekeeping, there is little sign of the bees being stressed by regular weekly inspections. The evidence for that can be seen and felt when the bees' famous 'bee dance' can often be observed continuing whilst a comb is being inspected outside the hive. It is also notable that during the inspections, there is rarely any sign of the bees' anger or defensiveness. The author often goes right through several hives, lifting out each brood frame in turn, without getting a single sting, with the manipulations carried out entirely with bare hands. The author often says, 'Now, bees, you be gentle with me and I will be gentle with you. If you get angry and start stinging in earnest, I will have to put gloves on, and then you won't like it!' Gloves are clumsy, and clumsiness annoys the bees, as do rapid movements. It is the author's view that bees become entirely used to the beekeeper's regular visits, which may account for their docile behaviour. In the author's experience, bees which are 'left alone' often become extremely bad-tempered whenever they are investigated. 'Left alone bees' are often the source of many swarms, most of which will probably die out without the care from a beekeeper.

The predominant hives in use in Britain are the National and the Smith followed by the double-walled W.B.C. with its attractive pitched roof. Then there are large hives, favoured by some beekeepers and much used in America such as the Dadant and the Langstroth. Another large hive, less common, is the Glen. The author's first hive was a picturesque, white-painted Cottage hive with a pitched roof. It took National size foundation on Smith frames. The choice of hive depends chiefly on what sort of beekeeping is adopted. If you want to move your bees to take advantage of different honey crops, you may be better off with Smith hives which take up the least amount of space in the back of a car. Again, if space saving is important where you keep your spare hive parts then the Smith is the best type of hive. The Smith also has a top bee space unlike the National which has a bottom bee space. That means that in the Smith hive, the bees can walk about more easily above the frames of the brood nest in the winter. In the National, the cover board is directly above the frames and movement of bees will be more restricted leading to a slightly greater risk of 'isolation starvation' when the bees become too cold, hungry and weak to search out where the stored honey can be found. Some people dislike the Smith because although it uses exactly the same size of foundation as the National, the frames have cut-down lugs at the top of the frame. The lugs rest on the runners or side rebate of the crate, and the frame is arguably easier to handle with long lugs found in the National frame. However, advocates of the Smith hive find that the short lugs are no problem, and will often handle the frames simply by holding the top bar anyway, using the short lugs simply as extra support. Some people dislike the Smith because it has a wide and a narrow side of the box or crate, the narrow side accommodating the short lugs of the top bar. Beekeepers with the Smith very quickly get used to that difference and simply find that having a narrow side is useful when the hive must be fitted into a small space. Whether it is Smith or National which is adopted, it is important to remember that although both types use exactly the same size of wax foundation, the frames of a National will not fit into a Smith hive even though a Smith frame can be put into a National hive, provided that precautions are taken if travelling to prevent the Smith frames moving about in transit.

The beekeeper should be aware of his or her limitations concerning bodily strength before adopting the larger types of hive such as the Glen, the Dadant or the Langstroth. Even a Smith or National super can be heavy to lift when it is full of honey and lifting a hive with supers on top, as when returning by car from a different site, can be a real strain. There is also the question of weather affecting yields, and in the British weather, it is possibly not very common to have the sustained good conditions that can justify a larger hive. Of course it is always fairly easy to put more boxes on top of a smaller type of hive in order to achieve a bigger colony of bees or more room for honey if the weather becomes ideal. Yield from bees is more dependant on good management than it is on the size of hive, and naturally to get lots of honey you need to have lots of bees. You can have lots of bees in the smaller types of hive as well as in hives with larger boxes. Once again, management, weather and other conditions are probably more important than the dimensions of the hive.

Single or more than one brood box?

Some beekeepers prefer to have more than one brood chamber because they can then rear more bees, and theoretically get more honey. The commonest pattern is one deep and one shallow crate for the brood, sometimes termed 'one and a half,' with the queen excluder and supers on top. Even though 'one and a half' is popular, and the author has seen large harvests of honey on hives arranged in such a way in the middle of London, single brood chamber management is widely practised too, and is the author's preference, in Scotland anyway. The argument for single brood chamber management is that in a cold climate the bees will make swarming preparations even before they have filled the single brood chamber, and any surplus honey will be put in the super because of limited space for the queen to lay in the single brood chamber. In Scotland, the main honey flow is often over by the time an additional brood chamber can be filled with brood, so that instead of concentrating on honey and swarm preparations, the bees will be intent on rearing more brood when those extra bees may no longer be of much use for collecting nectar with the honey flow over. When the author asked his mentor why he did not practise double brood chamber, or, alternatively, 'one and a half' brood chamber management he replied, 'if it was any more productive I would be doing it!' Single brood chamber management will probably induce earlier preparations for swarming, the hive being more crowded, but that can be an advantage because the bees can 'get over' their swarming stage earlier, and have a new young queen with more vigour to take advantage of the later honey flows such as heather. The early honey harvests can also be valuable with limited room to store it 'down below' in the brood chamber, and swarming preparations are, after all, the natural instinct of the honeybee to reproduce the species by setting up colonies in other places. 'It's just like a cow having a calf!' my mentor used to say. Whether you are going for single or double or 'one and a half' brood boxes, you will generally need at least two supers for honey storage in a reasonably good year, and an extra brood box, floor, crown board and roof are always useful for housing a swarm, even if the swarm is soon going to be united with your single colony. Swarms are very good at making new comb, and your extra brood box will be valuable even for that purpose.

It is sound practice to have a nucleus hive or box for each main hive which you have, even if you are not wanting to increase the number of colonies you own.

Assembling your hive.

Buying your hive un-assembled or 'in the flat', is certainly cheaper if you are buying new equipment. Galvanized oval nails and suitable glue are probably the best for holding the joints together, and putting together a hive delivered 'in the flat' does not present any problems even for people who claim that they are 'all thumbs'. Some people prefer to build their own hives from raw timber, but the idea has seldom attracted the author because the manufacturers have selected the best available wood, usually of western red cedar, which has some preservative qualities, is light and allows the wood to breathe a little. Hives must be built to very precise measurements. If you are building 'from the flat' remember to buy tinned rails and appropriate short nails of brass to fit the rails over the rebates upon which the frame lugs can rest. If rails of the correct length are not available, you may need a small hacksaw to cut them to the correct size to fit the box you are using. The author had to do that when building his own nucleus hives – a small hive – of which more will be said later.

Frames and bee-space.

Making your own frames is not often practised because the precise measurements and quality of timber are even more critical than is the case for building the hive itself. All the many frames must be very exactly made and all practically identical so that they can hold the foundation which is also of exact measurements. Foundation is a sheet of beeswax impressed with the hexagonal bee cell pattern. The bees build their comb upon that pattern which saves them a lot of extra work building wild comb which might not be straight and would not be reinforced with wire as it usually is for extra strength. There are many varieties of frames for sale by the beekeeping suppliers. The frames you buy must be suitable for the type of hive you are using. Many people nowadays use the Hoffman self spacing frames for the brood chamber which are available for all hive types. Similarly, the Manley closed side frames are generally chosen for the supers. The advantage of the closed side of the Manley super frames is that there is less risk of injury if you have to cut the combs out of the frame using a knife, or if you are using an 'uncapping knife'. There is also a reduced chance of the bees building 'brace' or wild comb from the side of the frame to the side of the super box. Previously, 'metal ends' were widely used to arrange the frames at the correct distance from each other in the hive. The correct distance apart is about a quarter of an inch, which is the bee-space. Any spaces in the hive of less than a bee space are likely to be attached to each other by the bees with brace comb. This brace comb can be an useful source of wax, but it is also often a nuisance when you need to move hive parts. If you leave spaces of more than a bee space in the hive, it is also possible that the bees will build wild comb in these areas, and they will then attach the wild comb to the crown board which will be a nuisance. With regard to the width of top bars of frames, there is also a variety on offer. The author prefers the wide top bars because there is then less likelihood of the bees building crooked combs or combs in the wrong place. If the frames have a narrow top bar, there is an increased chance that thy will build the combs too wide or build brace comb in between frames, making manipulation and movement of the combs more difficult.

Ekes.

The eke is a crate, often home-made, and used for honey production. The eke is of smaller dimensions than an ordinary shallow or super crate. Usually it is placed inside a super and two of its sides are rebated to take simple slats of wood which form top-bars. The eke thus resembles the construction of a top-bar hive in some respects. There are generally no side or bottom bars in an eke for honey collection, and the advantage is that the eke is very convenient for cut-comb honey production since the combs only need to be cut off the slats of wood at harvest. The slats are usually furnished with a groove to take a starter strip of thin foundation, 'glued' to the slat with a little melted beeswax. The slats or top-bars are retained in position with panel pins inserted through holes drilled in the sides of the rebates. The panel pins prevent the slats from moving about in transit. The eke is of a depth which is suitable for the size of cut-comb boxes which you are using. To prevent the bees from building their natural comb right down to the brood frames below, it is usual to place a sheet of heavy duty plastic on top of the queen excluder, the plastic being cut to about 5cm less than the size of the eke, so that the bees have access to the combs around the sides. Some beekeepers dispense with a queen excluder at this stage of the year and with the plastic sheet in place. Many beekeepers regard the use of ekes as rather old-fashioned, and simply use ordinary supers. Ekes are also used for varroa control using volatile oils such as is used in 'Apiguard' but are then specially designed for that purpose.

Assembly of frames and foundation.

This is a straightforward if fairly time consuming operation. Frame nails of the correct dimensions are required, the foundation should be at hand, and you may find it useful to have small panel pins to attach the strip of wood or 'wedge' to the top bar. Some people simply drive a frame nail right through the 'wedge' transversely but that may make dis-assembly more difficult if you need one day to fit new foundation in the frame. The bottom bars can also be secured with a frame nail right through transversely, but again, subsequent dis-assembly is likely to be difficult without destroying the frame. The author prefers to nail the bottom bars straight down vertically into the side bars even though nailing into the direction of the grain of the wood is not so strong, but in this instance it is probably good enough, especially when the bees will strengthen the joint with wax and propolis (the antiseptic bee glue from plant resins). Practice should be made with the correct use of a light hammer. The hammer movement should come only from the wrist, keeping the arm still, otherwise there is a real risk of breaking the foundation. The author prefers to work on a board placed on a chair opposite to the one he is sitting on. Stout folding chairs are used, and a chair is also useful during hive inspections so as to save one's back from injury from constant bending and lifting! Great care is needed when sliding the foundation into the grooves of the side bars of the frames. The frame is generally partly assembled only, until the foundation has been fed into it. Only the side bars and top bar are put together to begin with, and perhaps a single bottom bar, assuming you are using small twin bottom bars for brood frames, with the remaining bottom bar being nailed in when the foundation is in place. Side bars are nailed to the top bar with only two frame nails, one each side, hammered in transversely. It can be helpful to warm the sheet of foundation very gently in front of an electric fire or fan heater for a minute, or perhaps placed in a sun-heated car for a short while. Slight warming brings out the aroma of the wax, and liquefies the lighter elements making the foundation more translucent and less brittle. It may also make the foundation more attractive to the bees than cold, dry foundation. Before sliding in the foundation or attaching the side bars to the top of the frame, the strip of wood or 'wedge' which is generally still loosely attached, needs to be removed with a knife, or even your nails if they are strong enough. You are only likely to make the mistake once, of assembling a side bar the wrong way round, with the groove on the outside instead of on the inside of the frame!

A box of equipment for beekeeping.

In addition to your bee veil or bee suit, your gloves or elastic cuffs, you will require the following:

1. A strong hive tool.
2. A smoker.
3. Fuel for the smoker.
4. Manipulating cloths of canvas attached and rolled round dowels of wood about 1/2" thick and as wide as the hive is wide front to back.
5. A lighter or matches.
6. Water sprayer. with ordinary water. If the bees are boiling out of the entrance to attack, a quick spray of water may put them off! Also useful to cool off bees when travelling.

If travelling with bees:

7. Travelling screens should of course be in place instead of roof and crown board.
8. If travelling the roof and crown board must not be forgotten!
9. Drawing pins are useful for attaching
10. Bands of tape of strong fabric to fix the travelling screens to the super or brood box.
11. Hive straps, to hold the hive together.
12. Foam closures, to close the hive entrance.
13. A light hammer to tighten up lock-slides.
14. A screwdriver of medium size and length.
15. A decorator's scraper, to clean off brace comb.
16. Planks of wood and small blocks to keep hive off ground and level.
17. A hive carrier. The author uses one of the hinged wooden gripper type. This two-person operated carrier is very secure.

If collecting a swarm:

17. A large sheet to place under board and skep, capable of closing at the top for moving.
18. A board slightly larger than the skep.
19. A skep.
21. A large board for hiving the swarm where the hive is to be situated.
22. Newspaper large enough to cover the brood box should be handy if you are going to unite colonies.

How do I get my bees?

The author got his first bees, and many more, from an aged retired friend who 'had bees'. He can be called 'Sam' to hide his identity as he has living relatives. Sam never carried out hive inspections, and used to say, 'You would be far better to leave the bees alone instead of disturbing them so often!' My mentor always said, 'Don't listen to Sam, he knows nothing about beekeeping, why?, he hasn't been into some of his hives since before the Second World War!' Sam lived in a nearby cottage with no running water and a 'sit ooterie' the product of which, when well rotted on his midden certainly made his garden grow well. Sam had about a dozen hives in various states of dilapidation, with the bees often coming and going anywhere but hither and thither of the main entrance! Sam's bees populated the countryside with dozens of swarms which was a great help to the author who was able to start new hives with those swarms, or strengthen existing colonies with an influx of extra bees – always using the newspaper method for uniting colonies of course. A swarm, being temporarily broodless, should be given a trickle treatment with a very small amount of suitably diluted oxalic acid in water in order to control the varroa mite. See the section on treatment of disease, page 32 and pages 47–51.

Nowadays, with the serious reduction of bees and therefore also of swarms, starting your beekeeping with a swarm is not so easy even though it is the best way because swarms are champion builders of new comb. The bees have to build comb in the brood chamber first before they can be expected to put any honey up in the super for the beekeeper!

How do I catch a swarm?

A swarm is in many ways the best way to start your beekeeping. That is because you can treat them right away with the oxalic acid trickle treatment against the varroa mite. The oxalic acid treatment is only suitable for bees without larvae or at least very nearly so. Within a very few days, that opportunity will be lost as the queen in your swarm will start laying immediately if it is a 'top swarm', and soon after she gets mated if it is a 'cast swarm' (see section on swarming, chapter 3). As soon as you have seen your swarm and it is settled in a mass of bees on a branch, fence or wall, you must proceed with all reasonable haste because a swarm will not usually stay on a branch for very long. The scout bees will be out searching for what they perceive to be a good permanent home. As soon as they have found it, they will be off! If the bees do not look as though they are settled on the branch, they can sometimes be induced to settle by spraying only the bees in the air with a little clean water with not much force, 'rain', and by covering the branch on which some of the bees have settled with a large piece of cloth which will, perhaps, give the bees the idea of some shelter. Assuming the bees have settled a bit, the beekeeper takes his skep and places it directly under the swarm. With a firm shake of the branch the swarm is then shifted off the branch into the skep. The skep is then carried to the ground where you have placed a board resting on top of a large sheet. Very gently the skep is then inverted over the board and sheet, and a block of wood or a stone is placed gently under one edge of the skep. The bees on the ground will then crawl up into the skep, and attach themselves inside the top. As soon as the bees are settled a bit in the skep, it should be carried, closed with the sheet, to the empty hive and the skep is then shaken onto a large board, sloping upwards directly into the hive entrance. Very soon the scouts should induce all the bees on the board to proceed up the slope and into the hive. If your swarm is on a wall the procedure is a bit different. It is then necessary to pick up very delicately a handful of bees and shake them down in front of the entrance to your inverted skep or a spare hive if you have it handy. The bees remaining on the wall are then smoked off the wall with persistence, and the bees in the skep or hive will usually signal 'come in here' with the Nasenoff organ in their tails to the other bees still on the wall. The bees should not be left in the skep for more than an hour or two, because there is a strong risk that the swarm will take off again as soon as the scouts have found a 'better' home. It is thus advisable to hive the swarm as soon as most of the bees have gathered in the skep. A swarm can also be hived by shaking the skep directly into the top of the brood box, but most beekeepers prefer to see the mass of bees shaken onto a board, and then they can watch the splendid procession of bees up the sloping board and into the hive. Small puffs of smoke can be used to 'herd in' any stragglers at the edge of the mass of bees if they appear to be getting lost.

Many people today are starting with a nucleus stock of bees, generally on only five frames instead of a full hive. You should see that your nucleus is 'queen right' which can be ascertained by looking for eggs and developing larvae in the combs. Try to see if the brood looks healthy: there should be no sunken and discoloured brood cells and few, if any, dead larvae. A small amount of 'chalk brood', caused by a fungus, may be unavoidable in early spring. There should be no signs of dysentery or bees' diarrhoea on the combs, frames or at the entrance. If seen, it is often a sign of the nosema bacillus, and you should look for a healthy nucleus stock instead. If you get your nucleus stock as early as possible in the season, it will have a good chance of building up strength to see it through the following winter and draw out extra frames of foundation to make it a full hive. Unless there is a good 'honey flow' on, it is wise to feed the bees with one part sugar and two parts water syrup which will encourage the bees to build the combs and increase brood rearing. If possible, your nucleus stock of bees should have a young laying queen, preferably of the current year so that she will be a vigorous egg layer, but she should be no more than a year old for a quick increase in the strength of the colony. Anti-varroa medication should be given to the stock at the earliest opportunity. If the colony is broodless, as it will be just after a swarm has settled in its new home, you will have an excellent opportunity to use the safe and effective 3.5% oxalic treatment method which leaves no residue. Very soon, however, brood will be in the new colony, and other treatment methods will have to be used instead.

Stings

A common misconception is that swarming bees are fierce, and will sting anyone in their way. Nothing can be further from the truth. Swarming bees are generally full of food for their flight, and for setting up their new home. 'Full of food, empty of malice!' It is generally possible to walk right through a mass of swarming bees, and not receive a single sting! One can usually shake a swarm of bees off a branch with bare hands, and not receive a sting. Bees can be picked up by the handful, which creates a pleasant buzzing sensation in the hand, and if you are very gentle the bees will usually not sting! The bees in the hand can then be guided or shaken in front of the entrance to the skep or hive. If, on the other hand, the bees have been sitting on a branch for more than a few hours, they will be getting hungry, and their stores 'on board' will be used up and they could then be much more prone to stinging. The author once collected a swarm from a branch, shook it into the skep with bare hands, and got several stings on the hand below. The swarm had been there for several days! Remember that a few stings are usually good for you: bee sting therapy is very widespread in Russia, and the author's grandmother who kept bees about an hundred years ago was once severely stung all over her head. She had to spend a day or two in bed because she couldn't open her eyes properly on account of the swelling! Prior to the accident, she suffered from arthritis. After the stinging she never suffered from that disease again. She returned to beekeeping and lived on into her nineties! The author's mentor used to say, "Forget about stings and just get on with the job!" He never wore gloves, and unless the bees get angry, the author does all manipulations with bare hands. It seems probable that bees actually like to have contact with the skin of the beekeeper. Often they can be seen to lick up sweat! They also come to trust the beekeeper, and the beekeeper learns to trust them too. If you inspect your hive regularly, the bees very soon will get used to you and stinging will be a rarity. Always avoid crushing a bee, because that infuriates them. Handle them gently and they will reward you with quiet behaviour. Remember that gloves are clumsy, and if wearing them, you will probably no longer be treating your bees gently. If you get stung, scrape the sting off immediately with a scraper and blow smoke on the spot. The smoke will tend to neutralise the powerful scent of the sting, which acts as a signal to all the bees in the hive. After catching a swarm and your bees are a little settled in your hive, it is a good plan to feed them with 1:2 sugar and water if they have comb to build. You cannot expect the bees to produce honey for you until they have built their home in the brood chamber. When they have done that, and the queen is laying well, then they are able to start building comb above the queen excluder and filling the new super comb with honey for you!

Bees from commercial suppliers.

Bees can also be bought from the beekeeping suppliers as 'package bees' with a laying queen. The most obvious disadvantage of buying package bees is that they are expensive. The second drawback is that often we do not know where they have come from and there is therefore the problem of bringing bees into the country with we know not what diseases, and we are also without knowledge of their suitability for our climatic conditions. It was widely remarked that the yellowish Italian bees, although they were prolific in producing lots of brood were slow to put on much honey. It was also said that when the Italian bees crossed with the local black bees, their offspring were unusually bad-tempered. A similar and worse outcome was when the African bee was crossed with the European bee in S.America. The result was the infamous 'killer bee' which has become widespread in America although some claim that its fierceness is now a little attenuated. The novice beekeeper is strongly advised therefore to get his bees from a known local source, so that there is a good chance that the bees will be suitable for local conditions. Advice should be sought from your local beekeepers' association. There is a national association of beekeepers who are trying to increase the stock of our indigenous black bee. You may be able to find a source of black bees from BIBBA – the British Indigenous Black Bee Association. Starting with a nucleus populated with a good laying queen and her offspring in April or May is probably the best way of getting started if you have tried and failed to get a swarm in May or June the previous year, or a good nucleus stock the previous summer. The advantage of starting with a nucleus in April is that someone else has had the work of getting that small colony through the winter safely, and has also had the work and good luck to get that queen mated during the previous summer. There is also nothing wrong with buying a nucleus stock in the late summer when the queen can be seen to be laying well after getting mated. If the late summer nucleus only has an old queen from a previous year, she will probably be a little slower to build up a strong colony the following spring. Whatever sort of nucleus you have, remember that it must be treated for varroa, and it must be given plenty of sugar syrup, winter strength, to see it safely through the winter. Nucleus stocks require a disproportionately large supply of food to keep warm over the winter! A weak nucleus can also be rather prone to diseases such as nosema, which is often considered to be endemic (see chapter 8).

The bee castes.

When you have got your bees, it is important to be able to identify the different castes of bees in the hive: queen, drone and worker are the three castes. You must also be able to notice the difference in the three types of larvae or brood. Knowing these differences is important for the effective monitoring and management of your hives. Knowing what normal brood looks like will also help you to identify abnormal brood if your bees should be affected by disease or perhaps dysfunctional conditions.

Chapter Two:
Getting to know your bees and management of your hive of bees in the spring.

About the middle of April, when the days are becoming a little warmer and the flowering currant is out, the beekeeper needs to remove the mouseguard from the entrance of the hive. Spring feeding of two parts water and one part sugar should have been given from the middle of March onwards in order to encourage the rapid build up of the bee population. Without lots of bees you cannot get lots of honey. It should be remembered throughout the season that there is a considerable period of development which must elapse before our worker bees are ready for gathering nectar and the hive can produce honey. From the time the egg is laid by the queen, twenty one days must pass before the worker bee emerges from the cell. But before a worker flies forth from the hive she undertakes many other tasks first. At the outset she is generally busy with housekeeping: cleaning out the cells and polishing them ready to receive the queen's eggs. Then she is busy with nursing the young eggs and larvae. Soon she undertakes a variety of jobs about the hive, such as storing the nectar and evaporating the excess water from it and digesting the raw nectar to make honey. Pollen must be mixed with honey to form the food for nurse bees to give to the larvae. Wax must be created from the scales produced from the underside of the abdomen. The hive must be ventilated by fanning, and if it is very hot, water must be evaporated to cool the hive. The hive may need defending, and contrary to expectations, guard duties are reportedly performed mainly by young bees. When the worker is ready to fly from the hive, eighteen days have usually elapsed since she hatched, and even then she usually goes to collect water, pollen and propolis before she becomes a nectar collector. All these various tasks and stages mean that our worker bees cannot be expected to produce any honey for around six weeks after the eggs were laid. This suggests the importance of early stimulative feeding of the hive in order to build up a strong colony of bees in advance of the main honey flow.

Supering.

When the dandelions come out, and the bees are working at bringing in pollen and nectar, the feeders should be removed and the queen excluder should be placed directly over the frames of the brood chamber. Immediately on top of the queen excluder you place the 'super' or 'shallow' crate fitted with frames and foundation for extracted honey production or starter strips for cut comb management. Failure to super early enough will result in brace comb full of honey being built upon the crown board, and it will even encourage earlier swarming as a result of overcrowding. It is best not to give the bees more than one super to fill at a time, otherwise they will not fill them properly but will 'chimney' right up to the top super leaving empty frames at the sides. Place another empty super under the first super when that first super is three quarters full. The bees generally move up to the top of a box and build their combs downwards; comb is normally built by a cluster of bees hanging from the top bar of a frame. Always use a crown board on top of the frames of either brood box or super, depending on which box is on top. Failure to have a crown board under the roof will result in the bees building wild comb from the roof, and it may become almost impossible to remove the roof at the next inspection!

The bee castes.

When carrying out your first inspection of the brood chamber, you will be getting used to recognising the three different castes of bees. You should also be observant of the three types of brood corresponding to the three castes. Queen and worker eggs are fertilised from the sperm stored in the queen's spermatheca since she was mated. Drone eggs are not fertilised, so that drones have no father, but only carry the queen's inheritance (parthenogenesis). Worker cells have fairly flat cappings when sealed over, and they are laid in worker cells. Drone cells have convex cappings, and are laid in larger drone hexagons. Queen cells are laid in queen cups extending slightly outwards from the comb, with the entrance facing downwards. After about a week, usually eight days after the egg was laid, the queen cell will be lengthened out to about an inch long, hanging down from the face of the comb and with the tip sealed over. When that is seen, know that swarming is imminent and you are in danger of losing a large proportion of your bees unless swarm control measures are taken immediately! During the swarming season, it is important to be familiar with the stages of larval development from the tiny eggs up to the time when the cell is sealed (see chapter 3). Know also that if eggs are seen in worker cells, the hive probably has a laying queen, and is said to be 'queen-right' although there are occasional exceptions to this observation (see 'Swarm control summary', page 32).

Clip and mark queen.

Provided the weather is not too cold, about the middle of April, the chance should be taken to search for the queen, and when found, she should be clipped and marked. Some people are against clipping off half of one side of the queen's wings with sharp scissors, but the author has always done it because it is the kindest thing to do for the whole hive of bees, and the most productive thing for the beekeeper to do. Without a clipped queen the hive is very likely to catch the beekeeper out with a swarm of which he or she may be quite unaware except that suddenly, at the weekly inspection, the hive is found to have no eggs or queen in it, to have fewer bees than expected, and to have fully developed queen cells. The author has been caught out several times by unexpected swarms issuing from the hives even before the first queen cell was expected to be sealed. The bees are only supposed to swarm nine days after the egg was laid in a queen cell cup. The queen cell is supposed to be sealed on the eighth day after the egg was laid. A day or so later, the queen will lead off a swarm to found a new colony. That is how the bees reproduce the species. In hot weather, however, the bees may just say to themselves, 'We haven't read the books!' and then they will just swarm anyway!

Some people suggest that the bees may speed up the sealing of the first queen cell, but that is perhaps a little doubtful. Whatever actually happens, it is certain that if a swarm issues from the hive with an un-clipped queen, there is every likelihood that the bees will be lost, and set up a colony in some place where survival will be difficult or impossible. A swarm can also alight in a position which is a nuisance to others. If, on the other hand, the queen is clipped, the swarm will go forth, the queen will get lost but the bees will return to the hive, frustrated. If the beekeeper is lucky, the queen and a small cluster of bees may be found around her on the ground nearby, and the clipped queen and her entourage can be put into a nucleus box and an 'artificial swarm' can be set up in another hive or in a nucleus box.

Technique for queen marking.

Three or four gentle puffs of smoke should be given into the entrance of the hive and then the beekeeper removes the first two or three frames from the hive and places them, preferably in a nucleus box, after inspecting them for the presence of the queen. Each frame is then examined carefully using the manipulating cloths to expose only one frame at a time. As soon as the queen is seen, the cloths should be used to close off the top of the hive completely and extremely gently the frame with the queen on it can be rested on the cloth nearest to you. Approaching the queen from behind she is then picked up by the wings between the forefinger and thumb, and her legs are then offered to the middle finger of the opposite hand with the forefinger and thumb of the opposite hand closing very gently around her thorax. Some people use one of the queen holding gadgets available, but my mentor always warned me of the risk of damaging the queen with such gadgets. 'You can easily decapitate the queen with one of those things', he said. Queen marking and clipping should be carried out over a cardboard box because it is very easy to drop the queen and lose her in the grass! 'Environmental' correcting fluid can be used to place a single small drop of marker on the queen's thorax. Avoid using too much fluid because it can damage the queen and the hive will then reject her. As soon as she is successfully clipped and marked she should be returned to the comb and watched for a few seconds. If the workers cluster round her too much she and the workers should be given a few gentle puffs of smoke to prevent them 'balling' or smothering her on account of the strange smell she has picked up from you and the correcting fluid. The smoke will lessen the evidence of that strangeness and give the bees something else to think about. Very carefully the frame is then returned to the hive, which is then closed up as soon as possible after returning the outside frames to the brood box. Remember that queens practically never sting humans. However, if you tend to shake when carrying out these delicate operations, it does help to hold the wrist of the hand catching the queen, and to keep the elbows pressed into your sides. It also helps to rest the forearms on the cardboard box when clipping and marking. If you fail to find the queen, you should close up the hive and wait to find her at the next inspection. If you still fail to find her, you can remove all the frames from the hive one by one to another brood box, and the queen may then be found on the last frame, or on one of the frames you have already looked at. Some queens are good at hiding, and the clusters of bees can easily be moved aside or even removed from holes in the comb by touching the bees very gently with the palm and fingers of one hand, or a tiny puff of smoke can be used to clear the bees out of a hole in the comb. You should try your best to have the queen marked by the end of the first week in May, because swarm preparations often begin about then. With your queen clipped and marked there will be a considerable reduction of the beekeeper's anxiety about losing a swarm, and swarm control measures will be made much easier because they mostly involve finding the queen and making an artificial swarm of some sort before a natural swarm makes off with most of your honey surplus and most of your bees!

Chapter Three: Management of your hive.

As previously mentioned, it is natural for bees to swarm when conditions are suitable for increasing the species by setting up new colonies. However, the aim of the beekeeper should be to make an artificial swarm in order to avoid losing the bees and thus ensure a harvest of honey and the creation of new hives. As well as keeping your bees, you will want to avoid a natural swarm from your hive being a nuisance to neighbours in the bees' attempt to start a new colony in the neighbour's roof or garage!

Swarm control.

The first aid to swarm control is to have a marked queen and better still, a queen which is also clipped. When there are signs of swarm preparations in the hive, the beekeeper should take steps to forestall the swarm or at least take steps to ensure that the swarm is not lost. The beekeeper should not rejoice too much at the return of a swarm if the queen had been clipped. Clipping the queen only buys a little time. You could say that a clipped queen gives the beekeeper just a little latitude in the timing of hive management. A clipped queen helps to avoid the very real possibility of the beekeeper being 'caught out' and losing most of his or her bees in a swarm as well as the queen. With a clipped queen, she will get lost, but the bees will return to the hive without her, until they depart with a virgin queen about nine days later unless the beekeeper intervenes.

Weekly inspection of hives.

The first requirement of productive beekeeping is to carry out regular inspections of your hive. The author generally carries out a weekly inspection of the hives from the first week in May onwards. Some people who 'have' bees, but often do not 'keep' them will argue that it annoys the bees to go into their hive, and that the best thing for the bees and the beekeeper is to 'leave them alone'. The author has found that there is no truth in that view and that bees 'left alone' are usually extremely fierce when doing any manipulations such as taking off a super of honey, and going into the brood box of such a 'left alone' hive is likely to result in many stings for the beekeeper. From the author's many years of experience it is evident that bees get very used to the gentle manipulations of the beekeeper during weekly inspections. It is also best for the beekeeper to avoid using gloves because they are clumsy and that always annoys the bees. Usually the bees can be seen to crawl over the beekeeper's hands and fingers without any sign of aggression. The author has often seen them licking up sweat, probably for the minerals which it contains. The time to use gloves is only when the bees get bad tempered. Often the author goes right through a hive with bare hands without getting a single sting. 'Forget about stings, just get on with the job!' the author's mentor would sometimes shout or, 'I've got to make you more frightened of me than you are of the bees so that you will not hesitate to do what is required!'.

It is very important to keep an accurate record of whether queen cells are found when the weekly inspection of the hive is carried out. By knowing the timetable of queen cell development, the beekeeper can forestall the bees' determination to swarm and thus keep his hive united for honey gathering. The beekeeper may also allow them only an artificial swarm or perhaps a nucleus stock or two so that the number of hives can be increased or to provide a back-up if a hive fails to get a mated queen.

To begin the inspection, three or four moderate puffs of smoke should be given into the entrance of your hive, and a minute or so allowed for the bees to quieten down. The hive tool should then be inserted into one corner between the brood box and the super and a little smoke puffed in between the boxes before separating the boxes and lifting off the super and queen excluder. The super, temporarily removed, can be placed conveniently on the upturned roof of the hive, and the super can be covered with a piece of cloth to give some shelter to the bees and honeycomb within. It is helpful to remove temporarily the outside frame or two of the brood chamber when carrying out the weekly inspection. Before the outside frame is removed, the gap should be smoked to remove as many bees as possible so as to avoid crushing them, and the frame is then lifted very slowly with the same intention in mind. It can be leant against the hive very gently, or it can be placed in a spare nucleus box whilst manipulations are carried out, and then replaced at the end of the inspection when closing up the hive. Manipulating cloths are used to expose only one brood frame at a time, which keeps the hive calmer. Small puffs of smoke at the frame tops enable the beekeeper to prevent the bees from 'boiling up'. Smoking of bees should be skilful: 'not too much, not too little, but just right' to maintain control.

Each frame is then lifted up carefully and inspected for queen cell cups with eggs in them, or even larvae. People have usually allowed for an inspection every nine days because it takes eight days from the time the first egg is laid in a queen cell cup for the larva to develop into a sealed queen cell. The beekeeper should commit to memory: "Three days eggs, five days larvae, sealed on the eighth, hatches on the sixteenth." Some of these days are for the development of a queen only. Workers and drones take longer to hatch. As a rule, the hive will give off a swarm very soon after the first queen cell is sealed, the old queen going off with a large proportion of the hive, and the beekeeper being in danger of losing most of his or her honey-gathering workers. Because swarms can sometimes issue from the hive earlier than nine days since the last inspection, the author recommends a weekly inspection, which may extend to nine days if the weather is inclement. If the beekeeper does not mind losing an old and clipped queen, the inspection interval may be extended to a fortnight, because any virgin queen replacing a lost clipped queen could only hatch out on the sixteenth day after the first egg was laid in a queen cell cup and seven or eight days after the issue of a 'top' or first swarm and its return to the hive without the queen. It should be remembered that the hive will give off a cast or second swarm with a virgin queen about eight to nine days after the clipped queen was lost or alternatively removed from the hive, which means that further action must be taken to avoid losing most of your honey gathering stock!

At the weekly inspection, the beekeeper should look carefully at every comb containing eggs and developing larvae, and should examine any queen cell cups, which are small cup-like cells extending out from the face of the comb slightly. By turning the comb up into the light it is possible to see to the bottom of the cup, and if an egg is seen it should be noted in your notebook: 'egg in cup' and the date. The egg looks like a little white line or hyphen lying in the cell. If it is a week since your last inspection, it is quite possible that nearly, or actually, sealed queen cells will be observed. Queen cells are long cells extended from the cups and built out from the face of the comb, and have a dimpled appearance because they are made with a mixture of wax and pollen to enable good ventilation for the developing larvae (see page 37).

Queen cells.

If there are well developed queen cells in the hive, the eggs in cups having developed into larvae and fully formed cells since the last inspection, action should be taken immediately. The queen on her frame should be removed to another hive or nucleus box, and again the date she was removed should be recorded in your notebook. An artificial swarm of one kind or another should then be made or alternatively a nucleus box colony should be set up. If there is not yet a sealed queen cell, it is possible to estimate roughly when the cell will be sealed. By looking down into the open queen cell you will be able to see if the egg, noted a week earlier, has developed into the larval stage, and it is possible to estimate approximately what age the larva is. About three days after the egg was laid, it will be surrounded by a small amount of milky looking fluid and by the fourth or fifth day from when it was laid, the larva will surround the cell in a 'C' shape. When the larva completely fills the cell you can assume that sealing of the cell is not far off and a swarm will issue from the hive unless you do something about it!

Swarm control.

The usual practice is to make some sort of artificial swarm and to remove the queen from the hive. Removal of the queen merely delays the issue of a swarm, because soon after one of the queen cells has hatched, a virgin queen will lead off an 'after swarm' or 'cast'. By counting the days carefully from when the first egg was laid for a queen, you will be able to estimate the earliest date on which a virgin queen could lead off a cast swarm, i.e. sixteen days after the egg was laid. If several mature queen cells are sealed, and it is difficult to choose the best cell to continue the hive with a new queen till next year, recourse should be made to a careful examination of the cells in a good position. If the bees are working hard on a good, fully formed cell, choose that one, and mark it with a drawing pin on the top of the frame. A good position for a queen cell is in the middle of the frame and in the middle of the hive because that situation is warm and well covered by many bees and less likely to be damaged during manipulation of the frame. Often, you will need to touch the bees gently with the flat of your hand in order to move them aside so that you can see such cells. It is important to avoid shaking the bees off the frame which has the chosen cell on it. To do so could damage the young queen in her cell, or at least, separate her from her food supply, the royal jelly. The bees will therefore need to be moved about carefully with the flat of the hand so that no cells are missed when it comes to destroying unwanted queen cells and preventing a cast swarm. Because it is often very difficult to find all the queen cells in a strong hive, to get a better view, it is usually necessary to shake the bees off the frames into the hive when removing unwanted queen cells, always avoiding shaking the frame with the chosen cell as already mentioned.

Prevention of natural swarms.

A swarm is the way that honeybees reproduce the species by setting up a new colony in a place which the scout bees have chosen. 'Non-swarming bees' are a myth, and are not behaving in accordance with their natural instincts. 'A swarm is just like a cow having a calf' the author's mentor would say. Most beekeepers aim to prevent the issue of natural swarms from their hives by removing the old queen before the first queen cell in the hive is sealed. It must be emphasised that removing the queen only puts off the issue of a natural swarm for about eight or nine days, rarely for more, unless the weather is very bad. The author has frequently seen swarms go out even in cold wet weather, issuing from a friend's hive who 'had' bees but did not 'keep' them! The old queen, who leads off the 'top' or first swarm, may have been removed from the hive by the beekeeper, but the resourceful bees can simply use a newly hatched queen to lead off a 'cast' swarm a week to nine days later, as mentioned above. Six or seven days only, and no more, after the issue of a swarm, or alternatively, the removal of the queen by the beekeeper into another hive or nucleus box, you must go into the hive and systematically remove ALL the queen cells except ONE early, well-developed cell in a good position preferably in the middle of the frame and in the middle of the hive as mentioned above. Only if the beekeeper does that, the possibility of a cast swarm being thrown off will be almost eliminated. It must be remembered that failure to cut all cells except one will mean that the first queen to hatch is likely to lead off a cast swarm leaving the remaining queen cell or cells in the hive to head the remnants of your hive or lead off even more casts. A hive which has given off a large swarm and one or more casts is pretty useless for honey production for the rest of the season. Cutting out queen cells too early will encourage the bees to make many 'emergency queen cells' which are hard to find.

If the queen has hatched.

It may be that when you conduct your weekly inspection you will find that the good queen cell which you chose to continue the parent colony till the next season has already hatched. You should not be dismayed as long as a cast has not yet issued. Just write in your notebook with the date: 'Queen running', or 'hatched' and cut all cells. Sometimes, despite the beekeeper's best efforts, the hive will swarm anyway, and leave the hive queenless, but unless you see the swarm issue forth, that is hard to determine as there may be other causes of subsequent queenlessness. Assuming you are making a nucleus stock or an artificial swarm, it may be difficult or impossible to be sure that you have not shaken the hatched queen into the new hive or nucleus, with the result that you could have made the hive or parent colony queenless. Virgin queens are extremely difficult to find in a populous hive of bees, so it is probably best just to go ahead and make your nucleus or new hive with its complement of bees shaken off the frames from the parent colony. The author has occasionally introduced a newly hatched queen from the sealed cells removed from the parent colony, but that action has led to a swarm at least once. In order to reduce the chance of a queenless hive, it is sensible to have a 'back-up' plan even if you are not intending to increase the number of your hives. The simplest 'back-up' plan is to start a nucleus hive. Your nucleus hive can be headed by the old queen or, if you want a young new queen, you can aim to get her hatched from a queen cell removed from the parent hive on its own frame. The cell will hatch sixteen days from the egg, and with luck the queen will get mated and build up a new colony. Assuming your nucleus stock has only a fairly small population of bees, you can even leave several queen cells on the frame taken from the parent hive preparing to swarm. When the first queen hatches she will kill the other cells or queens and head the new colony. A nucleus hive, prepared the following way, being small and relatively weak, is very unlikely to throw a swarm.

The frame with the queen, alternatively queen cells, is placed in the middle of the nucleus hive, a frame of mostly sealed worker cells is included, as that will help to keep the bees in the nucleus busy, looking after the larvae, and forgetting about absconding. The hatched larvae will also help to populate the nucleus colony. You should also add at least two 'shakes' of bees to the nucleus box, and a frame of stores i.e. honey. Your nucleus will then consist of one frame of bees with the queen or queen cells, one frame with mostly sealed brood, and one frame of stores, plus two or three 'shakes' of bees plus sufficient frames of foundation or combs or dummy boards to fill up the box. If your nucleus stock is staying in the original apiary, you should stuff a small amount of long grass or weeds into the entrance. The idea is to make it a slight struggle for the bees to go in and out of the box. By the time the grass or weeds have withered the workers will have settled on the box as their new home. The bees notice their new surroundings and imprint its location on their memories which will reduce the 'drifting' back to the parent colony and prevent the nucleus being depleted excessively.

Queen rearing.

The simplest way to rear new queens is to set up several nucleus colonies instead of destroying, or allowing the bees to destroy, the excessive numbers of queen cells in the parent colony. In order to raise a new queen, the colony must want to raise one because it is queenless, or because it is around the height of summer, food is plentiful and the bees decide to swarm and reproduce the species. When the bees swarm, about half the colony goes away with the queen, leaving queen cells in the hive. There must be at least one queen cell in the colony (or even worker eggs must be present if the beekeeper is relying on the colony to make emergency queen cells). It is possible that queens raised on the 'emergency' principle are not as good as queens reared from normal large queen cells. After three to four weeks, the nucleus stocks should be inspected for worker eggs or brood, and if either are present it is evident that the new queen has hatched and been successfully mated. Queen rearing on a large scale is beyond the scope of this booklet.

Shaking bees into a new colony.

Holding a frame firmly by the lugs or top bar, the frame is given a single firm shake downwards over the nucleus box, with the result that most of the bees on the frame fall off into the box. If there are still a lot of bees on the frame, the shake can be repeated until practically all are dislodged. You can even thump the back of the hand holding the frame to dislodge remaining bees. The nucleus box should then be filled up with frames of foundation or combs and possibly dummy boards if you only want to give the bees one frame of foundation at a time to avoid the tendency for the bees to make a mess of drawing out new comb by nibbling the foundation or not drawing it out regularly.

With a marked queen, she can be easily found and removed on the frame on which she is situated and put into a nucleus box or another empty hive or even a matchbox if she is intended for somebody else. If you want to increase the number of hives you own, instead of placing the queen in a nucleus box as a 'back-up' queen, she can be placed in a new hive at a distance from the parent hive, and an artificial shaken swarm can be created. A frame of mostly sealed brood and also a frame of stores should be added and combs or foundation to fill up the brood chamber. Obviously, any queen cells must be removed if the queen is to go into a relatively well populated hive. Four or five 'shakes of bees' should be added to this artificial swarm hive, and grass can be stuffed loosely into the entrance to encourage the bees to notice that they are on a new site, and to mark its position for their return after flights. The only difference between setting up a nucleus hive and a full hive is the number of frames and, or, shakes of bees put into the new colony.

Treatment of the parent hive.

Six or seven days, no more, after the artificial shaken swarm was carried out, the parent colony from which the queen or queen cells was removed, should have all the queen cells removed except the marked one. If you do not remove the extra queen cells, the parent hive will throw off after swarms or 'casts', and the original hive will become hopelessly weakened with little or no honey. In the old days, many beekeepers used to kill the old queen, and cut out all the queen cells except one on the same day that they removed or killed the queen. That seems to the author to be a waste of a queen and the cause of more difficulty a week after the queen was removed and the swarm temporarily prevented. If the beekeeper cuts out queen cells too soon the hive will simply raise more queen cells from worker cells. In other words, the bees will convert ordinary worker cells into emergency queen cells. They do that by feeding a worker cell with an enriched diet and by building out the worker cell from the face of the comb. These emergency queen cells are much harder to find on the comb, and it will be very easy for the beekeeper to miss a cell so that the bees can still swarm with a virgin queen. The emergency cells are smaller than the usual queen cells, and have some similarity in appearance with a hooked nose and are sometimes called 'stubbies' To avoid emergency cells, it is best to cut out redundant queen cells after the age when worker eggs can be converted to queen larvae. Five days after the queen was taken out, the worker eggs should be too old for conversion to queens, but the author has occasionally had emergency cells built about five days after the queen was removed, so it is best to cut cells on the sixth or seventh day and before the ninth day. The frame with the chosen cell should be carefully checked for any other queen cells, by moving the bees gently with the flat of the hand. All other frames can have the bees shaken off into the hive if need be in order to inspect each frame thoroughly for queen cells which must be removed.

The disadvantage of the 'artificial shaken swarm' is that many bees will fly back to the parent colony unless you have an 'out apiary', that is an apiary which is more than two miles away. If your artificial swarm hive is in the same apiary as the parent hive, drifting of bees back to the parent colony may make it rather weak, and a weak hive is not good for gathering honey or for surviving the winter. The Pagden method of artificial swarm is better because it strengthens the swarm stock which is established on the original site of the hive including the old queen but no queen cells. The parent stock is moved away to a different place in the apiary, and having the brood and chosen queen cell, it should also have enough stores and pollen in it, as well as plentiful brood to prevent the parent colony from dwindling too much. Six or seven days after the queen cell was sealed in the parent colony all queen cells except the chosen one should be removed to prevent cast swarms. After two or three weeks, and a bit of luck, the chosen queen cell will hatch and she will get mated and start to lay, so that the parent colony's fate should be assured of success. By moving the parent colony to a different place fairly nearby many bees from it will drift back to the original site which has the old queen and is the 'swarm' stock. The bees drift because they tend to fly back to the original site which is retained in their memories. With this method, the artificial swarm stock on the original site is likely to collect plenty of nectar and yield a good honey harvest because it will have most of the flying bees 'drifted' from the parent colony. In order to drift even more bees and deplete the parent colony still more the beekeeper can use the 'Heddon' method, by which the old queen, on her frame plus a frame of mostly sealed brood and a frame of stores is placed on the original site with new combs or foundation to complete the hive as already described. The parent hive is then moved to the side with the entrance at right angles to the original entrance. After two or three days, the parent colony is then moved to the other side of the 'swarm' colony, and finally after a further interval, the parent hive is moved to the new site in the apiary. The effect is to 'drift' the bees by moving the parent hive first to one side, then to the other and finally to a new site (see pages 33 & 34).

Prevention of cast swarms.

Six or seven days after the 'Heddon' operation the parent colony should have all the queen cells removed except one in order to eliminate even the reduced risk of casts. It is sometimes suggested that because most of the flying bees will have flown to the original site where the old queen is situated, there should be inadequate bees for the parent hive to give off swarms. It should be noted, however, that virgin queens are often frisky and keen to swarm even with relatively small numbers, so no chances of losing bees should be allowed. However, a nucleus 'back up' colony, made with queen cells in it, and if it is in the original apiary, may be rather weak and is likely to suffer many bees making it back to the parent hive, and such a nucleus stock being relatively weak is very unlikely to throw off a cast, and therefore several queen cells can safely be left in it, the bees making the best choice of new queen. The same principle should not be relied upon if you have put many frames or shakes of bees into a full-sized hive, or if you are dealing with a still well-populated parent colony. If you feel that the parent colony is still very strong after shifting it with its entrance at right angles to the 'swarm' stock on the original site, you can always shift it again so that the entrance faces backwards and behind the 'swarm' stock hive which still has the old queen. After three shifts of the parent colony and then a fourth to the new site, the 'swarm' stock should be full of bees, and with few larvae and possibly no comb to build if you had spare combs, they should make a good honey gathering force, leaving the parent colony to raise and mate a new queen in the new colony. The principle of drifting bees to strengthen the 'swarm' colony is also fully employed with the Snelgrove method. This method of making an artificial swarm is usually practised using a Snelgrove board Having found the queen in the hive preparing to swarm, she is removed on her frame and placed in a new brood chamber as described above. Instead of placing the queen and a frame of mostly sealed brood and frame of stores in a new hive, she is merely placed in a new brood box on the floor of the original hive. All the brood frames are then placed on top of a Snelgrove board which is a crown board with the ventilation holes sealed up to avoid communication with the swarm stock below. The edge strips of the crown board are cut and pinned to make a hinged doorway so as to make new entrances for the bees from the edges of the crown board. The swarm stock with the queen is below with the supers and queen excluder above her but underneath the Snelgrove board. By opening an entrance at the side and at right angles to the 'swarm' stock, bees are drifted from the parent stock above to the swarm stock below. After a day or two, the entrance which the parent stock has been using is closed, and the entrance on the opposite side is used. After a day or two, an entrance at the back could also be used, and the beekeeper can go on drifting bees to the box below, the 'swarm' stock, as often as thought necessary. Once the parent colony is sufficiently weakened, it is very unlikely to swarm, but if the beekeeper feels that the bees have discovered his or her tricks, and are still powerful, the extra queen cells can be cut out as described previously. The advantage of the Snelgrove method is that less equipment is required and possibly less labour in cutting out queen cells. Less space is taken up in the apiary by this method, and the beekeeper does not have to move hives about in order to drift the bees to the 'swarm' stock. The disadvantage of the method is that there is more labour required to inspect the bottom box for swarming preparations, and when and if it is decided to increase the number of hives the extra equipment must be provided anyway. If you decide that you do not want extra hives, you can always unite the parent colony with the 'swarm' stock below when the parent has got a new queen mated. The old queen should first of all be removed from the bottom box before uniting, and the newspaper method of uniting colonies should be employed to ensure that there is no fighting between colonies.

When the bees are united you will have a number of frames too many to fit in one box, so you will need to shake the bees off redundant frames into the new hive. Frames with brood should of course go in the new hive, and a frame or two of pollen. Other frames can be kept away from wax moths for future use, but frames with unsealed honey can be fed outside to the bees, but not left out too long or they will chew through the combs and damage them in their anxiety to collect all the honey. Sealed honey may be extracted if you have an extractor able to take brood frames. Occasionally feeding outside can result in robbing, usually indicated by many bees struggling at the entrance of hives chucking out invaders from other hives. If you get robbing, the entrance to the robbed hive should be narrowed down until it stops. The author generally leaves wet frames and equipment out, and robbing rarely occurs, but there is probably more risk of it if other apiaries are nearby. Normally, the author regards brood frames filled with honey as belonging to the bees, and only honey from the supers is the 'rent' the bees pay to the beekeeper, but of course the Snelgrove system provides an exception to that general practice. The Demaree system of swarm control is rather like the Snelgrove system in that the swarm stock with the old queen is put into a new brood chamber on the floor. Otherwise, it is rather different because the brood, although it is put 'upstairs' above the queen excluder and supers, is not physically separated from the queen below except by the queen excluder and the distance through the super or supers. There are no upper entrances and no recourse to drifting of bees which was used in the Snelgrove method. All the queen cells are destroyed in the top brood chamber, and it must be checked for queen cells again after seven days. To avoid drones blocking the queen excluder and getting very frustrated, unable to go out, all the frames should be shaken into the bottom brood box with the old queen. The bottom brood box is filled with empty combs or foundation plus a frame of stores and a frame of brood without queen cells. The theory is that the queen, given plenty of room to lay, will not swarm. With the brood hatching out in the brood box above the supers, and the queen continuing to lay in the brood box below, the hive may become very powerful and will collect lots of nectar. The problem of this method is that since swarming is a natural process in the life of the hive, putting the queen in the bottom brood box with room to lay seems unlikely to frustrate the natural swarming instinct and there will be two boxes to inspect for queen cells, and much labour in lifting the top box and supers. You will also require an extractor capable of taking brood frames, to remove the honey from the top box after the brood has hatched out and the bees have filled the empty brood frames with honey. The theory is that the empty brood frames can then be given back to the floor of the hive with the old queen, and the procedure repeated with the new lot of brood on top again, above the supers and queen excluder. The system might work well if you have a continuous honey flow and continued fine weather, but these conditions are rarely present in Britain.

The author's preference is to use the Heddon system of swarm control if increase of stocks is desired. If no increase is wanted, the preferred method is to set up nucleus hives, at least one nucleus per hive, and to remove unwanted queen cells in the parent colonies six or seven days after the first queen cell was sealed. At the time of writing there is a strong demand from beginners wanting to start beekeeping, and it is recommended that unwanted nucleus hives be sold as soon as the new queens are successfully mated in the nucleus hive colonies. A nucleus colony can always be united with the swarm stock giving a boost to the strength of that colony if going to the heather.

If the queen could not be found.

If a hive is preparing to swarm and the queen could not be found, a 'last ditch' attempt to reduce the likely loss of bees in a swarm is to split the hive, putting five or six frames into each box. Queens are usually 'off the lay' prior to swarming, so you may not be lucky enough to infer the presence of the queen by finding eggs in one of the boxes. If you are lucky enough to be able to ascertain the box with the queen, you can adopt one of the swarm control systems described above.

If you do not want the old queen, she can be picked up and put into a matchbox with six workers to keep her warm and feed her. Workers are picked up in the same way as queens, and if care is taken to hold them by their wings, they will be unable to sting. The worker is put into the 'dark' end of the matchbox, opened just wide enough and quickly blanked off with the side of the index finger. If the box is left open a chink it can then be opened at the 'dark' end again to put in the next worker. When you have the six workers and the queen in the box, the situation can easily be checked by opening the matchbox against an opened car window. Next, about two millimetres of matchbox are opened and the narrow opening is stuffed with granulated honey from known healthy bees. Remember disease can spread with strange honey from colonies you do not know! It may be safe for human consumption, and dangerous for the bees! The best matchboxes to use are 'Swan' because they can be pushed into the wide entrance of many types of hives for queen introduction via matchbox. Kept in reasonably warm conditions, a queen will keep in a matchbox, topped up with honey, for a week, and she could be offered to other beekeepers perhaps for a lower price than that which is asked for a mated queen of the current year. Exchange of good strains of queens helps to avoid inbreeding.

Checking for queen laying and varroa treatment.

Queens take sixteen days to hatch after the egg was laid. Workers take twenty two days to hatch out, and drones take twenty five days. It is necessary to carry out varroa control with the appropriate 3.5% oxalic acid in a 20% sugar in water solution. The dosage is generally 30ml. trickled along the seams of bees for each hive. Treatment is given when the hive is without brood, or at least as free of brood as possible. Varroa treatment is therefore ideally given twenty six days after the queen was removed and all the worker and drone brood have hatched in the parent colony. The varroa mites leave the cell when the larvae hatch out and then attach themselves to the bees. In that situation they will be vulnerable to the oxalic acid solution. However, sometimes a queen will get mated and start laying before the twenty six days have passed, and in such an instance, treatment should be given as soon as eggs are seen in the hive again because the varroa mites will very soon start laying their eggs in the new open brood, shortly before sealing, and once the varroa eggs and bee larvae are in the cells, they are out of reach of the oxalic acid solution. The swarm stock should of course be treated against varroa soon, or at least as soon as the drifting of bees to it is past and before the old queen's eggs are nearing the sealing stage of the cells. As mentioned, the varroa mite generally migrates to the open cells soon before sealing on the eighth day. It has been recommended to the author that oxalic acid should be used only at an interval of more than three months since the previous oxalic treatment. Other methods of varroa control may need to be employed if the new queen in the parent colony starts laying early whilst there are still a lot of un-hatched cells in the hive. Formic acid pads are now becoming more available and will destroy the varroa even in the bees' brood cells. The synthetic pyrethroids such as Bayvarol and Apistan are effective, but the mite is becoming increasingly resistant to these chemicals, and they also leave small traces in the beeswax. The author generally uses the latter chemicals only after the honey has been taken off in the autumn, which minimises any residue in the wax because the wax is made in the summer. The chemical strips should be removed at the mid-winter oxalic treatment when the hive is practically broodless. Presumably, bees drifting from other apiaries may be carrying varroa mites, and if these mites are exposed to low levels of chemical they will develop a resistance to the medication, which is probably why the pyrethroid strips are only recommended to be left in for three months and no longer.

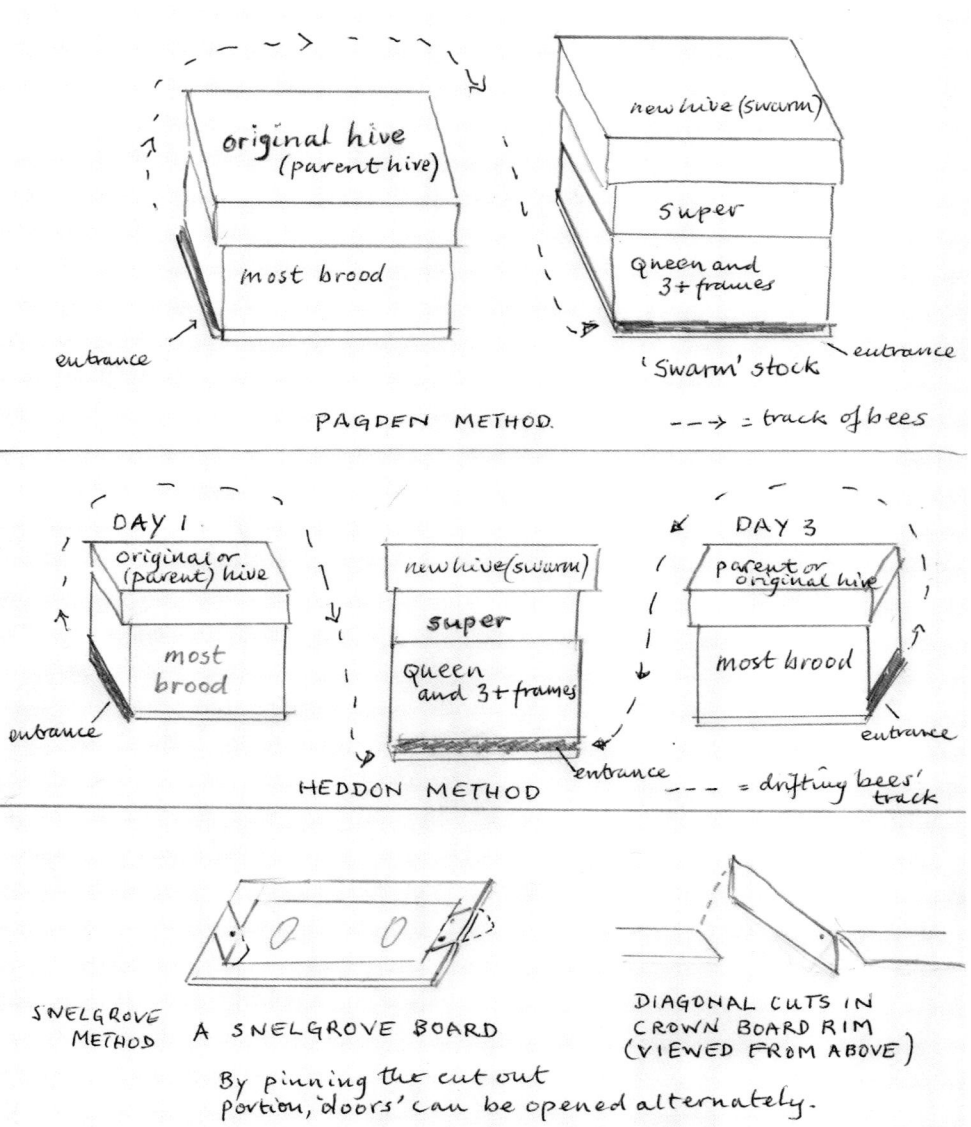

Queen cell.

Clipped and marked queen.

Drone cell and larva.

Worker with pollen basket.

Laying worker cells.

Laying workers. Note raised cappings in worker cells.

Queen cell cup.

Queen cell developed.

Sealed worker brood cells.

Sealed worker brood, and one sealed drone cell to the right.

Eggs in cells.

Larvae in cells.

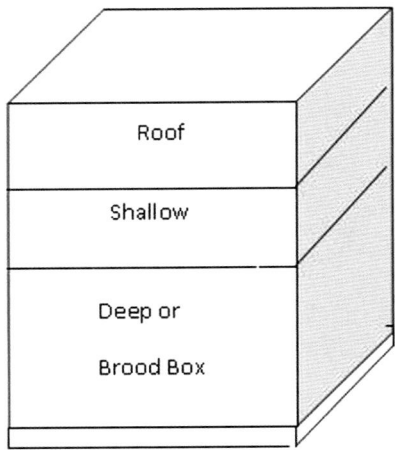

Diagram of typical hive arrangement for single brood chamber management. The queen excluder is placed directly above the brood chamber. With double brood box set up there will be two brood boxes, one above the other. With the 'one and a half' arrangement there will be two 'supers', the lower one being a small brood chamber, above which is placed the queen excluder and another super for honey.

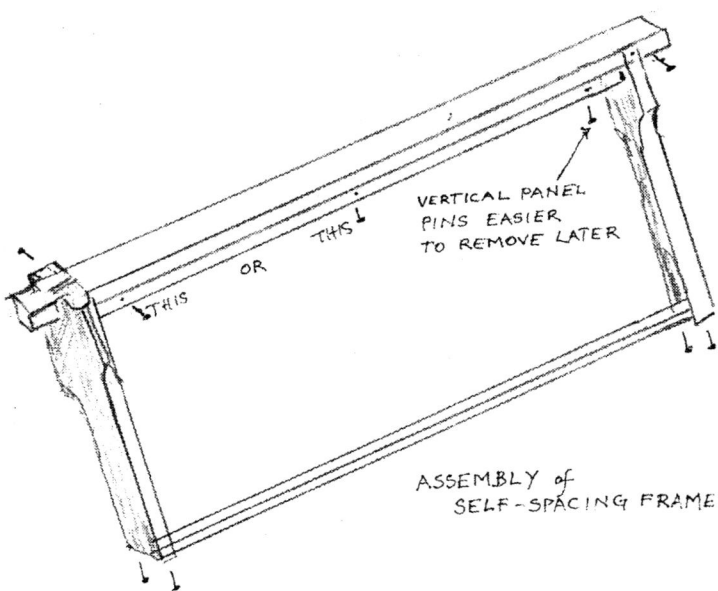

Swarm control summary.

This chapter has described several swarm control methods, and it is the beekeeper's choice to decide on the particular procedures used. It is probably helpful to understand that all the methods of swarm control employ three or possibly just two procedures.

1. The queen is marked before the swarming season, and preferably clipped as well as an insurance against getting the timing of a swarm slightly wrong.

2. When eggs are seen in queen cell cups, the date is marked in your notebook, and the hive is inspected for the cells' development one week later. Assuming there are developed queen cells at the latter inspection, the queen is removed and an artificial swarm is made. The date the queen was removed is noted down.

3. Six or seven days later, not more, all the queen cells except one are removed from the parent hive, leaving just one cell to continue the colony into the next season with a new queen, assuming she gets mated successfully.

Chapter Four: The Honey Harvest.

Most beekeepers are probably anxious to get their first honey of the season, and with regard to flavour, the honey straight out of the hive definitely has the edge on honey which has been around for a while. Honey which has been in the jar for many years even begins to taste just like commercial 'shop' honey of the cheaper varieties. There is nothing wrong with old honey, but probably some of the subtle aromatic oils and essences present in the original nectar combine with each other or with other substances in the honey, causing a change in taste.

The principal substances in honey are glucose and fructose. The glucose gives instant energy and does not need digestion. The fructose provides energy later on, making honey a very valuable food because there is not the peak and trough effect on blood glucose levels which comes with consuming sugar and glucose on its own. Honey provides sustained energy. That said, it is important not to eat too much honey at once because the glucose level in the blood can go too high and make you feel ill. The author once helped his mentor harvest the heather honey up in the moors. There were a number of broken combs and the honey was so delicious that excessive amounts were eaten there and then. The author felt very sick as a result, but having some medical knowledge, and by drinking lots of water from a nearby burn, thereby diluting the glucose concentration, recovery was almost immediate!

Taking off the honey.

It is probably a mistake to take the honey off too early. The shallow or super may feel very heavy, but if you take off the honey too soon, many cells may not be capped, and the honey may still have too much water in it. Honey with excessive amounts of water will not keep well, and will soon ferment. Too much water in the honey will probably contravene the trading regulations and will not be suitable for sale to the public. Even if the honey is ready to be taken off, it is said that leaving it on the hive encourages the bees to make more honey. The author generally removes the honey only at the end of the season, or if it is necessary to do so to make the hives ready for travelling. It is unwise to take the honey from the brood chamber. The bees need that honey to supply them with food when the weather is unfavourable or when they cannot collect nectar because good sources are unavailable. We should regard honey in the brood chamber as 'their honey'.

The usual way to remove the honey is to place 'bee escapes' underneath the supers. If you are intending to take the bees to another nectar source, and you are travelling with the bees, it is usual to clear the bees off the honey and into an empty super fitted with empty combs or with foundation or perhaps 'starter strips'. The queen excluder is, of course, placed under the new super. The traditional Porter escapes are fitted into the feed slots of the crown board after removing the gauze from the feed holes. The Porter escapes which the author uses are usually a fairly tight push fit into the feed slots, so the crown board may need supporting on a flat surface to avoid damage to the board. The Porter escape is fitted with the round hole side uppermost and the brass prongs on the underside facing onto the brood chamber or empty super. To avoid congestion of the bees, if clearing into the brood chamber, as you will be at the end of the season, it is best to remove the queen excluder from the hive. The queen excluder will not be needed until the following spring. The prongs of the Porter escape should be only about 1/8th of an inch wide or about 2mm. Care is needed to adjust the width of the prongs in order to avoid breaking them. The bees easily push aside the prongs, but find it hard to get back onto the honey. In July or August, the escape usually clears the super or supers of bees in twenty four hours. In colder weather in September, it is more usual to allow three days to clear. Care should be taken not to disturb the hive in any way when removing the cleared supers, and normally no smoke is needed, but should be on hand in case of any accident. Large sheets to cover the supers will help to avoid following bees as you take the supers away in a wheelbarrow. Any bees remaining in the supers can generally be removed with a long strand of grass to switch the bees out, away from the hives; they will find their way home directly. Care must be taken to ensure that all windows and doors are kept shut in the room you are using to process the honey harvest. The author's great aunt once left a window open in her honey room, and next day the bees had taken all the honey back again! It is astonishing how quickly bees can move even quite large volumes of honey!

Honey can also be removed from the hive by a more primitive method by shaking the bees off the frames. The author heard of one beekeeper who used that method, but she got lots of stings, and accused her bees of being bad tempered! The same lady apparently did not know about smoking the hive into the entrance before opening up the hive! Shaking the bees off their honey is obviously not a good idea, and it also carries the risk of breaking the combs and scattering the honey onto the bees which they will then have to clean up. In the old days, bees were sometimes cleared from supers by use of 'carbolic cloths' which had the effect of subduing the bees and made them run away from the chemical fumes. It is probable that this method is not often used nowadays because of the danger of tainting the honey with unpleasant chemical flavours.

There are a number of different designs of bee escapes to clear the honey from the combs, but the author has not tried them because the Porter escapes work perfectly well. It must be admitted, however, that the bees are inclined to glue up the prongs of the Porter escape with propolis, and if the delicate prongs cannot be cleaned with a fine screwdriver or knitting needle it may be necessary to dissolve the propolis out with real turpentine. The resulting dark brown turpentine can then be used with beeswax dissolved in it, to make a dark brown furniture polish, but the Porter escape must be well aired after it has been cleaned otherwise the bees could be deterred from using it. It takes a very long time to dissolve propolis in turpentine, and an equally long time to dissolve beeswax in 'turps' in order to make the best classic furniture polish. Recourse should not be made to any open flame or red heat on account of the serious fire risk.

Dealing with the honey.

If you live in an area where oil seed rape is grown within a few miles of your apiary, you are likely to have difficulty in extracting your honey, and it may be impossible to extract 'ripe' honey, which is honey with a suitably low water content. You can check to see if the honey is ripe, because if it is not, it will run out of any open cells when you tip the comb upside down and give it a shake. Honey should be dealt with within a week or two of taking it off the hive because honey is quite deliquescent. Honey which was ripe when you took it off the hive will very soon become watery again and may begin to ferment and will certainly be unsalable. If you are going to keep your honey in the supers for a week or two, it is best to shut off the air as much as possible. It can be very difficult to remove the combs from the super crate because the bees typically glue everything up with propolis and brace comb. The best thing to do is to turn the crate upside down and cut down with a sharp knife to separate the frames from the crate sides. You can then place the heel of the hand at each of the four corners of the crate and press down on the side bars of the frames with a lot of your own weight. That action will then generally enable you to push the frames out, and the super can be lifted clear of the frames.

Melting honey out of the combs.

Comb which has oil seed rape honey in it can be cut out of the frames and smashed up with a clean stick into metal or plastic buckets. The buckets can then be closed up until a convenient time to heat them slowly in a hot water bath which is thermostatically controlled to keep the honey and wax mixture below 140 Fahrenheit or 60 deg. Celsius. It may take as much as twenty-four hours to melt most of the granulated honey out of the broken combs. Avoid heating the mixture too quickly. The wax is skimmed off the top of the honey using a perforated skimmer, the wax being put into another bucket. If the mixture is allowed to get too hot, the honey will be slightly spoilt and may have to be used for cooking or mead production depending on how much the flavour is impaired. Heather honey, being higher in proteins, is much more easily spoilt by overheating than blossom honey.

The bucket with the wax in it can still yield quite a lot of honey, but it will have to be re-heated, if possible to a slightly higher temperature than previously so that the wax is almost melting on the top. Two clean sticks may then be inserted into the hardening wax, and before the honey underneath can cool too much, the honey can be streamed out through the holes made by the sticks into a sieve over another bucket. As the wax begins to break up, a large dinner plate can be used to hold it back so that most of the honey can be run out from under the wax. The remaining wax can then be washed with hot water if you live in a soft water area, and the washing water is then suitably diluted to make mead. A heated bucket with tap would make the procedure much easier, but such containers are very expensive.

When the wax is no longer tasting sweet it can be given a final wash in clean rain or distilled water, or water melted from ice in the fridge caused by condensation. The wax can then be drained for several days in a large sieve and finally dried out over kitchen towels until such time as you are ready to clean it and use it (see chapter nine, page 60).

Some people may succeed in extracting oil seed rape honey by heating the combs enough to liquefy the honey but not melt the wax. The author's attempts with a home-made comb heater were not very successful and honey melted out onto the heating elements. Also, the honey in the combs quickly cooled down with the result that this viscous type of honey was still difficult to extract even in the liquid state. Rape honey which is runny is likely to have a water content which is too high with the disadvantages already mentioned.

Use of an extractor.

Extracting your honey using a centrifugal extractor is the most productive and economical form of honey harvesting. Using wired foundation, the combs can be used year after year with the result that the bees are saved a lot of honey in the work of building new comb every season. Before the frames full of honey can be placed in the extractor, the combs must be 'uncapped' using a special long uncapping knife, either cold, with a serrated edge, or heated according to the beekeeper's choice. The knife is inserted just under the surface of the sealed honey cells, and the cut is often made upwards, with the comb slanted towards the uncapping tray beneath, so that the sheet of cappings falls onto the tray. The cappings may then be washed in soft water, or fed back to the bees to clean up. With the extractor loaded up, the revolutions should be increased very slowly to avoid damaging the combs. When most of the honey is out, the speed can safely be increased. The honey is run off into a settling tank, sometimes erroneously called a 'ripener', with a honey tap at the base. A honey sieve placed above the settling tank removes the fragments of wax. Honey should not be filtered because that would remove the pollen which has therapeutic value and probably enhances the flavour. Remember that honey flows silently!

Controlled granulation.

All pure honey will granulate eventually. The glucose and fructose simply crystallises out of the saturated solution which is the physical nature of honey. Some honey granulates very quickly, and oil seed rape honey is often found granulated whilst it is still on the hive! Many people find natural granulation makes the honey less easily eaten. Often the honey sets rock hard in the honey jar. To avoid that, a very little finely granulated honey can be introduced to the bucket in which the honey is settled or stored temporarily. Some people even introduce a little fine glucose, which is a natural component of honey anyway. The effect of 'seeding' the honey in this way is to cause it to granulate finely. If the honey is then stirred from time to time during the granulation process of several days, the honey never sets really hard, and this form of preparation is often termed 'creamed honey' and is reputed to sell well. Remember that even if the honey in your jar has become rock hard with natural, uncontrolled granulation, the honey can always be brought back to the liquid state by heating the jar a little in a hot water bath. The lid should first be loosened a little to allow expansion, and the bottom of the jar should be separated from contact with the base of the pan if placed on the cooker.

Cut-comb production.

Selling or using honey as 'cut-comb' avoids some complications, and certainly cut comb commands a higher price and is very popular if it is good. As previously mentioned, comb which has honey in it which is not 'ripe' is unsuitable. Honey which is granulated in the combs is also unappealing and is not likely to sell well. Oil seed rape honey is fine when mixed well with other blossom honey, and may be less beset with these problems than pure rape honey. Some people even dislike pure rape honey saying that it has a 'cabbagey taste'. Producing cut comb is made much easier if you have a 'cutting box'. This is easily made with four rectangles of white wood screwed together to make a long narrow box, open at top and bottom, with wires stretched across the top to cut the combs at suitable widths to fit into the commercially sold plastic boxes with clear lids. The comb is simply dropped gently into the cutting box which is then lifted clear and the pieces of cut comb are then placed onto a griddle to drain. They can then be put into the plastic boxes and the lid is snapped shut, keeping the honey dry.

Heather honey, like oil seed rape honey, is very viscous and does not extract easily. Some people use loosening needles to stir up the honey in the cells after uncapping, the honey being 'thixotropic' and machines and hand operated honey 'looseners' are sold by the suppliers. The author tried the hand looseners and found them not very successful and gave up extracting heather honey, preferring to make cut comb heather honey which is always much appreciated. Heather honey is also sometimes extracted by packing the combs into strong canvas bags and crushing the bags of comb which are inserted between wire griddles and pressed in a honey press. Once again, because of the difficulty in keeping the honey warm enough, but not spoilt by over-heating, getting the honey into jars may present problems because of the viscosity of the honey.

Selling honey.

Your beekeeping costs can be quite high, so selling some of your honey is a useful way of recouping some or all of your expenses. One of the best ways to sell your honey is to friends because you can usually get a higher price without having to sell your honey at wholesale prices to a shop. Even if you only have a few hives, friends and relatives may be unable to buy enough of your surplus honey, so you will have to telephone around the retailers to discover which shops will take your honey. Many retailers will work on a 30% mark up, and at the time of writing it is common for the public to have to pay £5 or more for a 1lb. jar of blossom honey, and considerably more for heather honey. Cut-comb heather honey often retails at £9 per 1lb. or more. Remember that unless you are able to extract honey from wired combs, it takes several pounds weight of honey to make each pound weight of wax, and the bees will have to build more combs next year if you are selling cut-comb honey. By the time you have prepared your honey for sale, you will realise how much work is involved in your beekeeping, and how important it is to ask a realistic price. Much of the honey sold in supermarkets comes from warmer countries overseas with lower costs of living than in the U.K. but with higher yields of honey per hive than is general here. When pricing your honey, it is therefore important to take into account the prices for local U.K. honey. The National Bee Unit publishes information on honey prices from its surveys of beekeepers, and the author has used its online data to assess prices pertaining for honey.

Keeping to the regulations.

It is easy to fall foul of regulations governing the sale of food when selling honey. Most beekeepers are amateurs and are small scale producers, so the regulations may not be as elaborate as those governing manufacturing, nevertheless, rules must be followed to avoid trouble. Honey must show the nett weight in grams, the country of origin, and the name and postcode of the beekeeper and a 'best before' date. It is also probable that the beekeeper will be required to be registered with the local council as a food producer. It is probably helpful in such an instance to speak to the council official or write to the official and explain your situation because any attempts by officials to treat the amateur beekeeper as a food manufacturer should be resisted in order to avoid possible requirements to have the beekeeper's premises brought up to industrial food production standards, which could be very expensive and 'not worth the candle'! As mentioned previously, honey should be fit for sale with regard to water content, extremely clean and suitably sieved to remove wax but not filtered. If honey were to be filtered, it is thought that this would remove some of its health-giving properties because the pollens can be not only nutritious but can also confer immunity from adverse reactions to air-borne pollens, which occurs in hay fever.

Chapter Five: The June Gap.

After the fruit blossom is over, and early oil seed rape has faded there is often a dearth of nectar coming into the hive, and nucleus hives and artificial swarm stocks have often used up all the honey stores supplied to them. Particularly if the weather is bad, and because of high winds and rain, the bees are unable to forage. It is often vitally important to give sugar syrup in such circumstances. Very often also, the beekeeper is expecting the bees to make new comb, and it should be remembered that to draw new comb, the bees require plenty of food. They will quite happily draw excellent comb if you feed them on 1:2 sugar syrup, and feeding in such a way will also induce continuing brood rearing during the June Gap if you have a laying queen. Failure to feed a colony in its infancy is to expose it to the danger of starvation. Do not use brown sugar because it can cause dysentery. Although sugar is not so good as honey because it does not contain the full spectrum of nutrients needed by the bees, it is closely similar to nectar in many respects, and nutrients are also provided by pollen when in season or in their stores. The bees also digest the sugar and turn it into invert sugar which is more suitable for their needs and is quite good enough to keep them going until nectar is again available.

Checking your hives after the swarming stage.

About three weeks after the swarm control measures were taken, it is vitally important to ensure that the queen cell left in the parent colony has hatched and the virgin queen has successfully been mated. You can tell at once whether mating has been successful because eggs and perhaps even early stage larvae will be found in the hive. It is usually un-necessary to inspect the hive sooner than three weeks after swarm prevention measures. The ideal time for oxalic treatment for varroa control would be twenty six days after the queen was removed from the parent hive. The reason for that, is that twenty five days are required for all the drones to hatch out. When the drones have hatched the varroa mites will all be 'hitching a ride' on the bees and will therefore be susceptible to the oxalic acid solution. When the mites are still in the cells, they are out of reach of the oxalic treatment. The mites favour drone cells, probably because they offer more food over a longer period than they would get from queen and worker brood. Some people advocate drone cell destruction as a varroa treatment, but the author rarely uses that method because it is probably bad for hive morale, it reduces the number of drones available to mate with queens, and also, the varroa can inhabit worker cells quite easily. If the new queen is found to be laying only three weeks after the old queen was removed from the hive, oxalic treatment for varroa infestation should be given without much delay because the mites will migrate to the cells soon before they are sealed on the eighth day from the egg, and when cells are sealed the mites are out of reach, as already stated.

The 'swarm' stock should have been given the oxalic treatment soon after the artificial swarm was carried out, no more than five or six days afterwards, before the old queen's eggs in the new colony have had time to be sealed as mentioned on page 24.

Although formic acid treatment is effective against varroa in the cells, it is more trouble and less pleasant to administer than oxalic acid, and it may be less acceptable for the bees. Synthetic chemical strips are best used at the end of the season, and may be declining in effectiveness as previously mentioned. Until resistance to pyrethroids becomes a significant problem, the author will continue to use a combination of carefully timed oxalic treatment and apistan or bayvarol strips in the autumn. The ultimate solution to the varroa problem is surely the development of natural resistance by the bees. The form this will take will almost certainly be special behaviour patterns adopted by resistant strains of bees, described as 'hygienic behaviour'. Colonies have already been identified showing this behaviour, and efforts are being made to propagate more varroa resistant colonies. Until these strains of bees become more generally available, it is likely that most beekeepers will have to continue relying on chemical or physical treatments such as dusting the bees with caster sugar.

If your hive is queenless.

After a month, and if no eggs or brood are found in the hive, it is fairly safe to assume that the hive is queenless. The hive should therefore be given a laying queen very soon, usually by uniting with another queen-right colony. She may have got lost on the mating flight, the hive may have swarmed and left the hive queenless, which they do occasionally, or the virgin queen may even have died. A queenless hive will often have a number of 'dud' queen cells, which will hatch drones or are completely without viable life. An easy test for queenlessness is to remove a frame from the hive and replace it with a frame with worker eggs from another hive after shaking off the 'foreign' bees to prevent fighting (see pages 8 & 49). Within four or five days, some of the worker cells will be drawn out into emergency queen cells (page 28) which confirms that the colony is queenless. Whatever the cause of queenlessness, the beekeeper must take action to prevent the extinction of the whole hive, and the difficulty of laying workers. If you have left the hive without attention for more than a month, laying workers may well have started. Their evidence is a large number of worker cells with markedly raised cappings like drone cells but smaller. The cappings have the appearance of .22 rifle bullets and these 'bullet cells' will only hatch into drones if they hatch at all (see page 36). Evidently some workers manage to develop an egg-laying propensity in the emergency of having no queen in the hive, but having no sperm, the eggs are infertile and cannot develop into a queen or a worker. A hive with laying workers is in danger of refusing an introduced queen, and must be given appropriate treatment. The treatment is to remove the colony about a hundred yards away and shake all the bees off the combs onto the grass. The laying workers are unable to fly back to the original hive site when you put the colony back where it belongs in the apiary. The 'good' workers will find their way back without trouble.

Bees' attachment to location.

At this point it is useful to recall that bees practically always act like homing pigeons: they are attached to their location, and if you move a hive, the bees will, for some time, return to the precise location where the hive was previously. This behaviour pattern has already been mentioned in the context of 'drifting bees' when carrying out artificial swarms. To avoid bees getting 'lost' it is usual to move a hive more than two miles away, or two feet or less per day. If more than two miles, the bees will register the new site as 'home'. If more than about two feet from their original site the bees will return to the original site unless special precautions are taken. To move a hive more than two feet away, it should be moved gradually, two feet at a time over a number of days and in the evening so that the bees are all back home and will recognise the changed position next day. Grass can also be stuffed into the entrance loosely to make the bees register that their situation is different. The presence of some brood in the hive will also discourage desertion by the bees as mentioned in the context of swarm control and setting up of nucleus hives.

Saving the hive with laying workers and uniting of colonies.

The beekeeper must make the hive 'queen right' once more. The usual way to do that is to move a colony with a queen next to the queenless hive observing the rule mentioned in the last paragraph. The brood box with the queen is placed usually under the brood box of the queenless colony, but a large sheet of newspaper is put over the underneath, queen-right, brood box. Small slits, not holes, can be made in the newspaper to speed up uniting a little, but it can still take a day for the bees to 'make friends after they have read the papers!' The newspaper method of uniting has not been known to fail in the author's experience, and when the bees have chewed away large holes the queenless stock bees can safely be shaken down into the queen right colony and uniting is complete. The combs from the formerly queenless stock can be kept away from wax moths for use at another time. If colonies are put together without the precaution of putting newspaper between the colonies, there is a high risk that the colonies will fight, and hundreds or thousands of bees will perish, stung to death.

Queen introduction.

A queenless colony can also be given a queen, but since she is a 'stranger' she will be killed unless precautions are taken. The new queen can be put into a queen cage of some sort, access to the queen usually made by workers having to chew their way through granulated honey first. The simplest method is probably matchbox introduction. As explained on page 31 a Swan matchbox, being shallow is ideal because it can be pushed into the 'wide' setting of the hive entrance. By the time the workers accompanying the queen and those in the hive have chewed away the slit opening of the matchbox they will allow the queen to come out and join the hive without struggle. It should be noted, however, that in cold conditions, and with a weak queenless colony, the queen in the matchbox may get chilled and she may not be successfully released from her 'prison'. The author once had that happen, and by breathing on the queen in the palm of the hand she was brought back to life and simply led into the entrance without any precautions and became accepted. By placing the matchbox above the brood box, or perhaps between the frames, chilling of the queen could be avoided. The queen is caught in the same way as mentioned on page 21 or sometimes she may be a virgin queen and may be released from a mature queen cell on the sixteenth day after the egg was laid. After the queen is put into the matchbox, she must be accompanied by six workers. Workers are picked up by the wings between forefinger and thumb, but the beekeeper must be quick! By opening the matchbox just wide enough to admit the worker, the bee is quickly pushed in and immediately the side of the finger is placed over the gap to prevent the queen or any workers already in the box from escaping before the box can be closed. Leaving a very small chink of light to enter, the workers and queen in the box are attracted to it so that the box can be opened quickly at the other end to admit another worker until you have six and the queen inside. By opening the box against a car window or piece of glass you can check that all is well. A very small gap is left in the matchbox, and the gap is stuffed with some granulated honey to keep the bees fed. The matchbox should be kept in a slightly warm place, and the supply of honey should be checked regularly. If that is done, a queen will usually keep for a week in the matchbox, but not for much longer.

Chapter Six: Travelling with Bees.

When you first get your bees, it is quite likely that you will have to pick them up by car and travel with them back home. Travelling with bees is also required if you are going to take them to new areas where they can forage for new sources of nectar. For example, when the fruit blossom and sycamore bloom is over, there may be a dearth of nectar secreting plants in your vicinity, and there is then the option of taking the bees to where there will be plenty for them to collect. As has been mentioned earlier, oil seed rape honey is easily collected, and some farmers may even be willing to pay the beekeeper to obtain better pollination. There is a risk, however, because many farmers are using systemic insecticides some of which may be harmful to bees such as the synthetic nicotine poisons like amidacloprid. Some of these pesticides have already been banned in Europe because of widespread damage to bees, but are still available in Britain. There is also the difficulty of managing the rape honey as mentioned in chapter 4. Oil seed rape honey is a good honey if mixed with other honeys but some people are not enthusiastic about pure oil seed rape honey, and it granulates very quickly because of the high proportion of glucose in it.

Native small-flowered white clovers, and not the imported large flowered varieties, have traditionally been an excellent source of good honey, and clover honey still attracts a slightly higher price than mixed 'blossom honey'. One of the most highly esteemed honeys is lime blossom, and it certainly has a wonderful flavour. Probably the finest tasting honey is heather honey. Both bell heather and ling heather have superb scent and taste and are well worth travelling for. Heather honey always commands a considerably higher price than blossom honey, and the price is fully justified for all the extra work involved. Collecting heather nectar is also fairly wearing on the bees, and going to the heather is not worth while unless you have a strong colony, if possible with a queen of the current year to replace all the bees 'worn out' by the hard work of foraging at the heather.

Preparing to travel.

Some of what follows is listed already under the equipment list, but this section may be read also for the techniques and cautions described. Preparation for travelling needs to be undertaken several days in advance. The bees need to be cleared off their honey, into an empty super fitted with thin foundation if you are intending to make cut-comb honey. The earlier honey removed must be placed in a secure, bee-proof and dry place until you are ready to process it (see chapter 4, taking off the honey). The empty super will give the bees the opportunity to start building comb and filling it with honey as soon as they arrive at the new place. It is unwise to travel with new, fragile combs containing new honey because these combs are liable to break in transit. This has happened once or twice to the author and to friends, and it was first supposed that the bees had overheated with panic on the journey, or perhaps that the bees had become 'car sick'. Bees can overheat on a journey, and it does no harm to spray the hive with a little water through the travelling screens perhaps half way through the journey. On the occasions when the sad accidents happened, and when the entrance closure bungs were removed at the new site, drowning bees, strong smelling fluid and honey poured out of the entrance. Broken combs were seen in the super, and when returning a day or two later to investigate, it was evident that the hive was dead. Consideration of the accidents indicates the probability that the weight of clustering bees on unsupported combs in the super is the major cause. When the combs fall and cover the bees with honey the bees probably panic because they are already in some stress with the journey and being shut in. With panic, they probably regurgitate the contents of their stomachs or may even defecate and the unpleasant smelling fluid issuing from the entrance is witness to their abnormal and fatal behaviour. When returning from the new site, perhaps in the autumn, the situation is different because the combs are then well supported and the temperatures then pertaining are usually lower. Some beekeepers remove the honey from the hives at the new location before returning the bees back home. The problem with that is the cost in time and fuel, because a journey will have to be made to the site several times. The author generally brings the hives home in the autumn with supers attached with the result that only two journeys are essential.

Immediate preparations.

A day or perhaps two days before departure, the new super should be installed over the queen excluder, and the crown board with bee escapes inserted is placed over the empty super. Above the crown board, which is now a 'clearing board' the supers full of honey are placed, and a bee proof roof is placed over the top. A day, or two days later, the bees should have cleared off the honey, which can be lifted off and taken to the bee-proof place where you can deal with it later. To prepare the hives for travel, it is necessary to have screen boards to give the bees ventilation at the top, above the empty super. The screens are left open, without crown board or roof, the latter are only put back on after the bungs or foam entrance closures have been removed at the new location, and the travelling screen is replaced with the crown board in situ again. It is a mistake to leave Porter escapes and screen boards on the hive for any longer than is absolutely necessary to prevent the bees from making a mess of them with brace comb and propolis. The floor of the hive should be secured with lock-slides, to which a couple of tightening taps may be given with a light hammer after you have just closed up the hive. The day before travelling, the crown board should be removed and replaced with a travelling screen above the empty super fitted with frames and foundation or starter strips. The crown board can be placed askew over the screen and likewise the roof to allow some overhead ventilation during the night, yet some protection from rain as well. Travelling straps are then threaded under and above the hive ready for tightening up prior to lifting the following day. At dusk, the entrance is gently but persistently smoked at a distance of a few feet, not into the entrance, but in such a way as to make it smoky immediately outside the hive which has the effect of driving the bees back into the hive if they are still hanging about the entrance even though they may have stopped flying for the day. As soon as the bees are all in, the plastic foam bungs are swiftly pushed right along the entrance, sealing off the bees for the journey. Next day the hive straps are tightened up, which holds the hive together and the hive is lifted with the hive carrier and taken to the car. It helps to have large sheets of cardboard in the back of the car so that the hives can be slid easily into suitable positions. If any bees are left behind, they will usually sit quite serenely on top of the travelling screens communicating with their sisters below. Bees do not usually fly about inside the car unless they are escaping from the hive in large numbers, in which case an opened back window will often suck them out to join another beekeeper's colony perhaps. If there is a leak, you must stop immediately and rectify the situation. It is useful to carry tacks, muslin or other sealing materials. It is comforting to remember that bees leaking out of the hive will probably not be vicious. The author has had a hive come apart when unloading, and although his hands were covered with bees he did not receive one sting. It is helpful to secure the screen with stout strips of tape and drawing pins because the screen is easily displaced when manoeuvring the hive straps into position. Before starting, ensure that all your equipment is on board and do not forget the crown boards and roofs! Planks of rough wood and small blocks of wood should be taken to the new site to keep the hives level and off the damp earth, and a sheet of some material can be placed in front of the hives to prevent excessive growth of herbage which could obstruct the entrance. When all is ready at the new site, and the hives are facing roughly south or south-east, the bungs can be gently removed. When the bees have settled down a little, the screens are removed and the crown boards and roofs are replaced back on the hives.

When returning home, the screens are generally left on until it is light the following day, but the bungs are removed gently even if it is dark so that the bees can get out early the following morning and normal ventilation from the hive entrance is again restored.

As soon as the screens have been removed, the crown boards are fitted with the Porter or other suitable escapes, the queen excluders are removed and the crown boards with escapes are placed on top of the brood frames (see page 42). The honey supers are then placed on top of the crown boards and the roof is put back on top. It is important to ensure that the roof is bee proof, otherwise the honey is likely to be robbed out! If there is brace comb with honey in it, it can be scraped off with the hive tool, and the honey fed to the bees at the entrance to flavour their winter stores. As soon as the bees are cleared from the supers the honey should be removed to a safe place, and if it is autumn the autumn varroa treatment should be given without delay and the autumn feeding started immediately. The autumn varroa treatment has to be some kind of treatment which is fairly long-term when brood is still in the hives such as Apistan or Bayvarol if there is not yet resistance to these treatments. If there is resistance, Apiguard or Formic Acid or Thymol pads will have to be used as soon as possible before cold weather makes them a less effective treatment choice.

In order to ensure proper winter preparations it is wise to return from the heather during the first or second week in September. Usually the main flourish of heather is going over by then, and the days are getting shorter, so that very little extra honey can be expected by staying much longer, and the advantages of early preparations for winter may be missed.

Chapter Seven: Preparing for Winter.

With the harvest of honey safely off, treatment for Varroa control and autumn feeding must proceed without delay. An empty super, without frames is placed above the crown board to cover the feeder. The author generally gives the bees four kilos of sugar per hive to see them through the winter until March – do not use brown sugar which can cause dysentery. The feed is given with the sugar dissolved in about the same quantity of water, i.e. 1:1. Hot water is added to the sugar in a pan, and the water is filled to a little above the level of the sugar and stirred until the sugar is completely dissolved. By the time the sugar has dissolved, the feed should be just warm. The most convenient feeders are those which take about one kilo of dissolved sugar. The bees easily take down and store one kilo overnight! The feeders have a central cone which is open at the bottom and allows direct access of the bees through the crown board feed hole. The bees crawl up the inside of the cone to the top which is above the level of the syrup. They then crawl down the outside of the cone to reach the surface of the syrup. The cone is covered with an upturned cup which reaches down almost to the base of the feeder. Sometimes the bees have been known to push the cup off its attachments with the result that some bees have drowned in the syrup, the bottom of the cup no longer reaching nearly to the top of the cone. By weighing down the cup with a piece of slate on top, drowned bees can be avoided and the bees would not be strong enough to lift cup and slate. As soon as the bees have nearly emptied the feeder they should be given the next kilo of sugar as syrup. The aim should be to get all the feed taken down and processed by the bees before the cold weather sets in.

A very basic feeder which is home-made can be made with a catering size coffee tin with a well fitting press-on lid. The tin is filled with syrup in the usual way, but the lid is punched with perhaps a dozen small holes from the underside and then pressed firmly onto the coffee tin. The full tin is then inverted directly onto the brood frames, and the syrup simply drips very slowly onto the frames below because it is held back by air pressure.

Before the cold weather arrives, mouseguards must be fitted to the entrance of the hives otherwise mice will get in and destroy the colony by eating the bees, larvae, wax and stores. With very gentle smoking to avoid excessive curiosity from the bees, two small strips of wood about a quarter of an inch thick are pushed into the entrance one at each end of the entrance to act as distance pieces. Perforated zinc, a strip of old queen excluder or a strip of wood is then pinned right along the entrance leaving the new entrance only a quarter of an inch high. The result is that the bees have access in and out, but the gap is just too narrow to allow the mice to get in. With the mouseguard fitted, the strips of wood are then gently removed.

If the hive roofs are fairly light, winter gales could blow them off, so it is sensible to put a brick on top of the roof. With all these precautions taken the beekeeper can be fairly confident that the bees are well prepared for winter.

If there has been a heavy snowfall, it is probably wise to see that the entrances to the hives are clear, otherwise there is the possibility of the bees suffocating, although it is also possible that the snow will still allow enough air to get into the hives. One common occurrence is the exit of numerous bees possibly attracted by the light produced by snow. Very frequently the bees settle on the snow and then become chilled so that they are unable to get up and then they die on the snow. Some beekeepers shade the hive entrances, or may even strew material on the ground in an attempt to reduce the white glare. The author has tried warming the chilled bees in the palm of the hand, but has found that sometimes they do not readily re-enter the hive even when picked up and put on the entrance board. These warmed bees have simply flown back and settled on the snow again. The question arises as to whether the colony uses the snow to allow old bees to commit suicide peacefully. It is a fact that many bees do die off during the winter, and sometimes mortality is so high that the beekeeper needs to use a piece of grass or small stick to keep the entrance clear of numerous dead bees which the housekeeper bees have not had the opportunity to clear away.

Occasionally woodpeckers have been known to chop through the wall of a hive and eat the bees during the winter. A friend's hives and colonies were destroyed in that way. The defence against that occurrence is to wrap the hives in fine chicken wire. The problem is probably more apparent in thin-walled hives and in areas close to woodland.

In the autumn, immediately the honey is off the hive, varroa control must be applied. Because there will probably still be brood in the hive, the treatment must be one which attacks the mites in the larvae of the bees as well as on the bees themselves. Synthetic pyrethroids such as Apistan and Bayvarol have been used for years, but now there is growing resistance to these substances, and formic acid pads or alternatively Apiguard which uses strong aromatic plant oils must be used instead. The problem with these latter treatments is that a fairly high atmospheric temperature is required to achieve the necessary diffusion of the substance around the hive interior, so that the treatment may not be entirely effective. It is therefore very necessary to kill as many mites as possible in the middle of winter during a mild spell using oxalic acid. It is said that oxalic treatment should not be given at closer intervals than three months, the bees becoming too acidic with more frequent treatments. The author's preference is for the trickle method, along the seams of bees using a small squeezy bottle (see page 32). In the middle of winter there will be very little brood in the hive, so practically all the mites will be attached to the bees and will therefore be susceptible to the oxalic acid.

Bees can easily starve as soon as brood rearing begins in March. About the middle of March the bees should be given spring syrup which is half as strong as autumn syrup, i.e. about 500gms. sugar to 1ltr. of water. The author generally removes the feeders as soon as the bees have taken down the allocated amount. Leaving the feeders on the hive will usually result in the bees covering them with propolis and making them less efficient or even unusable without a lot of scraping. Very little smoke is required to remove or put on feeders, but it is wise to follow my mentor's advice: 'never be without smoke in an apiary!'

Chapter Eight: Pests and Diseases.

One prominent and widespread disease and its treatments has already been mentioned, and the reader is referred to the chapter on Checking the Hive in Late Summer (chapter 5) and to the chapter on Preparing for Winter (chapter 7). The disease mentioned is of course, varroa, caused by the mite varroa destructor, which came into Britain in the late 1980s and early 1990s All the author's hives were killed not once but twice, probably as a result of varroa infestation and secondary virus infections. The problem with varroa is that although there may be only low levels of infestation, the mite evidently spreads very lethal viruses which then result in the rapid collapse of colonies. Although the author's colonies were treated with oxalic acid when nearly broodless, it is probable that some mites survived, and the viruses at those times were particularly lethal. Feral colonies were also seen to be eliminated at the same time in the vicinity. It is probable that the beekeeper must treat the hives against varroa at least twice a year, and any swarms collected should be treated immediately with oxalic acid solution before the queen begins to lay. The reader should recall that oxalic acid is ineffective against varroa inside the larval cells of the bees. Beekeeping suppliers are now selling a variety of varroa controlling systems. The author has been advised that treatment with oxalic acid should not be used too frequently, and a gap of three months or more has been suggested.

After the viruses spread by the varroa mite, the most serious threat to colony survival are probably the brood diseases. The most dangerous brood diseases are American Foul Brood, and European Foul Brood. Both of these diseases are notifiable, which means that as soon as either of them has been identified in any of your hives, the Government Bee Inspectorate must be informed. If you think your bees may have one of these diseases, an appointment for a government bee inspector to visit your apiary should be made without delay. A very useful illustrated booklet *Foul Brood Diseases of Bees* is available, possibly free of charge, from HMSO or probably direct from DEFRA. It is sufficient in the context of the present work merely to outline the main features of these diseases, as the reader is referred to the government publications.

American Foul Brood.

The larval cells will be seen to be darkened and discoloured and the brood can be seen to be dead and lying at the bottom of the cell. American Foul Brood is the more lethal and hard to treat infection of the two foul brood diseases. It is distinguished from European Foul Brood most easily by the 'matchstick test'. In American Foul Brood, a matchstick applied to remove the dead larva from a cell will show 'ropiness.' That means that when the matchstick is withdrawn slowly, it will bring with it a string of glutinous material from the cell. Eventually the dead larvae dry up in the cell, leaving a dark brown scale. Many cells may have jagged perforations in the sealed cappings as the worker bees attempt to remove the dead larvae. Brood will typically show a peppered appearance with many cells unsealed or not occupied with living larvae. Treatment of the disease is not generally reliable, and the only way to avoid the spread of infection is to destroy the colony, and to sterilise the hive parts with a strong scorching with a good blowlamp. All combs must be burnt.

European Foul Brood.

This disease is also notifiable, and treatment of the affected colonies needs to be supervised by a government bee inspector who will decide the appropriate action which must be taken. Light infections are sometimes treated with approved antibiotics, but weakened colonies and those with a high degree of infection are generally destroyed. Diagnosis is less easy than with American Foul Brood because the brood abnormalities can be confused with other brood ailments. Inspectors will generally send samples to laboratories for confirmation of infections, or will use a lateral flow device employing immuno-assay enzymes on site.

In European foul brood the larvae often die before the cell is sealed, and the larvae have a twisted, abnormal attitude in the cell. The dead larvae collapse and turn brown leaving a sunken brown scale which is not ropy when removed with a matchstick. Capped cells are sunken and discoloured. It may be possible to see the colour of the larval gut in infected individual cells. The colour is creamy yellow.

When the infection is fairly slight, an effective treatment is to separate the bees from infected combs by burning the old combs and re-hiving the bees on clean comb or foundation in a clean hive by means of the 'shook swarm' (see pages 26 & 27). The DEFRA booklet recommends sterilisation of equipment using a strong solution of washing soda because microbes are sometimes inadvertently transferred to other colonies by the beekeeper. For the same reason, bees should not be given honey from an unknown source which could be infected.

Chalkbrood.

This is a common condition, usually noticed in the spring, and seems to be related to cold and damp conditions. Brood cells are white and chalky looking, and the infection is fungal. The infection is said to affect only sealed brood. Comb replacement and removal of heavily infected combs is advisable to avoid spreading the spores. Chalkbrood is less of a problem with strong colonies.

Nosema.

This is a fairly common disease of adult bees, and is often troublesome in the early spring in weak colonies. There will often be signs of dysentery affecting the hive, with faeces voided at the hive entrance and even on the combs within the hive. A bad infection with nosema will often kill the hive. A slight or moderate infection can be treated by giving the bees clean combs if the combs are sullied and spraying the combs and adhering bees with an antibiotic known as Fumidil B which was obtained from the beekeeping suppliers. The author has read that this remedy has now been withdrawn. The infected colony should be placed in a replacement clean hive, and the infected hive is sterilized by placing a pad soaked in glacial acetic acid over the top bars of the frames, and the hive then sealed up for a week. The sterilized hive and saved combs should then be allowed to air until it ceases to be reeking of acetic acid. Rubber gloves should be worn when dealing with glacial acetic acid, and the operations should be carried out in the open air to avoid the fumes.

Nosema is diagnosed by means of microscopy using the 500X or 1000X lens. A negative stain is used, known as Nigrosin with the eggs of the microbe showing up white against the dark background. The eggs have the form of a somewhat flattened oval, rather like the oval round station names on the London Underground, but less elongated. A sample of dead bees is taken from the hive floor or hive entrance and the bees are placed in a mortar and mixed with a little water and then well pulverized with the pestle and mortar. A dropper pipette is used to take some of the pulverized bee mixture to a saucer, and a drop of Nigrosin is then added. A drop of the blackish mixture is then transferred to the microscope slide in the normal way for examination.

Paralysis.

This disease is indicated when the adult bees take on an unusually shiny black appearance and are seen to tremble a lot on the alighting board at the hive entrance. The bees appear reluctant or unable to fly. The disease may be viral, but it would appear to have a genetic susceptibility because the condition is generally cured by removing the old queen and giving the colony another queen, following the usual precautions when introducing a new queen. Since the colony will probably be weakened, it may be advisable to unite it with another colony using the 'newspaper method'.

Pesticides.

Recent research, mainly in America, has indicated that bees are exposed to a considerable number of agricultural pesticides, and that the precise effects of these chemicals are often not clear, and the possible interactions of several pesticides in the bees' bodies are unknown. There has been much anxiety about pesticides in beekeeping circles internationally, and some widely used neo-nicotinoids have been banned on the continent but are still used in Britain. Particularly worrying is the use of systemic insecticides on plants which are important pollen and nectar sources for bees. There is some evidence that insecticides such as Imidacloprid and clothianidin may not always be fatal for the adult bee, but that the insecticide-polluted pollen and nectar may have deleterious effects on larval development and subsequently on the adult bee too. Considerable discussion of Colony Collapse Disorder which has been serious especially in America has led to widespread concern, not only amongst beekeepers but also in the public at large because we are all so dependant on bees for pollination of so many food crops. At the time of writing, the cause of the large scale and widespread death of colonies remains a mystery.

If you suddenly see a large number of dead and dying bees in front of your hives, it is advisable to send a sample of twenty-five or fifty dead bees to SASA (Scottish Advisory Service for Agriculture) or your appropriate DEFRA agency so that they can test the bees for pesticide poisoning.

Acarine.

This is a mite infestation of adult bees. Its presence in the hive may be suggested by the evidence of bees on the alighting board with the two wings on each side separated instead of together as they are normally. The appearance is often described as 'K' wings, and the causative agent has been described as a virus, possibly spread by the acarine mite. The mites can be seen under the microscope occupying the tracheal tubes or spiracles of the adult bees. The trachea can be examined by placing a dead bee on its back and slicing through its body just behind the first pair of legs. The scalpel or razor blade is delicately pushed towards the head of the bee, removing the front part of the thorax. Using needles and a little water, the trachea can be teased out of the thorax and examined with a hand lens. The tracheal tubes may then be removed to a microscope slide and examined with the 50X or 100X lens. If no microscope is available, the trachea are usually dirty looking if acarine is present, and white if the bee was healthy. To obtain a sample of bees for any diagnostic purpose, an open matchbox can be closed carefully over a few bees, and the bees are then placed in the freezer compartment of the fridge for a day. Strips for the control of acarine are available from beekeepers' suppliers, but nowadays the disease is sometimes regarded as endemic and may cause little harm to a strong colony. Probably bees are becoming increasingly resistant to the mites. Some chemical treatments for varroa are also claimed to be effective for controlling acarine.

Braula.

This is a small reddish insect which feeds off honey from the bees mouth. It is not particularly troublesome to the bees, but it spoils the appearance of combs by tunneling just underneath the cappings of the combs, and making them less attractive for sale for cut-comb honey. If you destroy your honeycombs anyway because of the prevalence of oil-seed rape in your area, there will generally be very little if any braula present. Most anti-varroa treatments kill the braula, so it may be rare now except in feral colonies.

Sac brood.

This is described as a malady of sealed brood in early summer. The cell contents are said to be granular and watery and the larva can be removed in one piece. The disease is not usually said to be persistent.

Small Hive Beetle.

At the time of writing this pest has not yet been found in Britain, but it is probable that it soon will arrive. It is about one third the size of a worker bee, and eats bee larvae, eggs and honey. The S.H.B. larvae have spines on their backs and small legs near the head end. Pupation is in the soil, which provides a means of destroying the pest. The DEFRA scientific agency should be contacted if S.H.B. is found.

Tropilaelaps.

This is a mite pest which, to date, is not yet present in the U.K. but is also likely to arrive sooner or later. If found, contact DEFRA scientific agency, SASA in Scotland. The mites are described as about 1mm. in length. They only feed on brood, so are susceptible to pesticides applied at broodless periods in the beekeeping year.

Chapter Nine: Dealing with beeswax.

The first requirement is to remove all the honey from your wax. The cappings or broken combs floating on the top of your honey bucket must be carefully skimmed off. The resulting mush of honey and wax can then be washed in soft water only, perhaps after running off the honey still remaining from under the wax (chapter Four). The honey dissolved in the soft water can then be kept for mead making, and the reader is referred to books on home winemaking and brewing because that is a craft in itself. When the washing water is no longer sweet, the wax can be drained in a large sieve for several days, turning occasionally to dry it out. On no account should wax be allowed to go down the drain, especially if it is mixed with hot water. Any particles of wax floating on the top of hot water can be discarded on the ground outside, or soaked up in kitchen cloth to make fire lighters. The wax can be finally dried by placing on kitchen towels, and it can then be stored in a dry state for future use.

Your wax will need additional cleaning for it to be suitable for candle and cosmetic making, and perhaps also for making polish. Wax can be melted by placing it in very hot soft water, or better still, rainwater or distilled water, or water gathered from melted ice from the fridge. All melting of wax must be carried out away from any open flame or red hot plates because of the serious risk of fire. It is usual to melt wax in soft or distilled water in a container which is placed in a boiler or water bath of some sort. The wax floats up to the surface of the water and sets there in a cake upon cooling. A stick can be placed in the wax before it is quite hard to allow the water underneath to be drained off. When cool the cake can be removed and the dirty wax at the top and bottom of the cake can be scraped off and used for dark polish or for lighting fires when melted and soaked into rags of vegetable origin. One technique is to put the wax into fine cotton muslin bags with very clean stones to sink the bag to the bottom of your heated container of soft water. The muslin will filter out some of the dirty constituents of the wax, and the clean wax will float to the surface of the water. Wax can be further cleaned in the dry state by placing in an electric oven on a low heat and filtering it when poured hot through fine muslin. A frame of some sort is used across which to stretch the muslin to provide your filter. When the muslin becomes too dirty, it can be used for lighting fires as mentioned above for the first stage of cleaning. A solar wax collector is used by many beekeepers, and it has the advantage of working without the use of any other heating agent. A filtration or sieving system is generally included in the design, and the solar collector needs to be well insulated, and is often double-glazed or covered with U-pvc. Many beekeepers simply clean their wax and then sell it to the beekeeping suppliers in exchange for new foundation or other beekeeping supplies. However, there are several crafts for using beeswax, and there are other hive products worthy of mention here.

A brief mention of hive products and crafts related to them.

Candle making, the manufacture of cosmetics and furniture polish are crafts in their own right, and the reader should consult books dealing with these matters specifically.

Some beekeepers collect propolis which is becoming more marketable, and you can also make your own therapeutic tinctures by dissolving it in gin, but be careful to ensure that you are not allergic to the propolis, which will vary according to the plant resins the bees have been using in its creation.

Royal jelly is collected by some beekeepers, and there is a considerable market for it, but its collection requires the inducement of a great many queen cells by specialised treatment of the colony, and it has been observed that royal jelly production is quite stressful for the hive, and requires exceptionally strong colonies.

The production of large numbers of queens for sale is also a considerable business, is also rather stressful for a colony, and requires techniques outside the scope of this book. For the general good of beekeeping, the author considers that rearing lots of queens should preferably be undertaken by those wishing to confer particularly beneficial qualities on the bee population as a whole, such as hygienic anti-varroa behaviour.

Pollen collection is adopted by some beekeepers, and the pollen is then sold for therapeutic use. The collection of pollen should only be adopted if there are sufficient stores of pollen for the use of the hive.

Bee venom is collected in Russia using special rubberised steel plates. The bees are induced to sting by delivering a minute electric shock, and the venom is subsequently scraped off the metal plate. The bees survive the stinging in this instance, and probably live to sting again! The author has sometimes used bee sting therapy on himself, and has sometimes offered it to others, with few takers! By placing a worker bee in a matchbox, the box can be opened on an arthritic joint. A sharp smack on the matchbox will generally cause the bee inside to sting.

Glossary.

Apiary: A place where hives of bees are kept.

Cappings: When the bees have evaporated sufficient water from the honey, they preserve it by sealing the cells with cappings of wax. These cappings must be removed if it is desired to run the honey out by extraction.

Caste: There are three castes of honeybees: queen, worker and drone.

Cast: A cast swarm is a swarm following a 'top' swarm, generally eight days or a little more after the laying queen and her entourage have departed for a new home. See 'swarm' below.

Colony. This is simply a hive or Stock of bees. It is usual to speak of wild or feral hives as 'colonies'. During the swarming season, the terms hive, stock and colony are used more frequently to distinguish a parent hive giving off a swarm, and its offspring, which is usually a nucleus stock or a new hive.

Cutting Box: A box, open top and bottom, with wire stretched across the top for cutting combs of honey in suitable sized pieces.

Cut Comb Production: Instead of using wired shallow foundation, bees draw natural comb which is then cut into pieces of required size.

Deep: Refers to brood frames and foundation sizes, also to brood box.

Drawn Comb: The result of bees turning foundation into combs.

Drifting bees: Describes the tendency of some bees to move naturally from one colony to another. It also describes the manipulation by which larger numbers of bees can be persuaded to join another colony during swarm control operations.

Dummy Boards: These are generally made of plywood attached to a frame space top bar. They are used to fill up space in a hive instead of frames and foundation. They may also be used to isolate part of a hive, but the floor arrangement and space above the top bar must be checked to ensure that isolation is effective. They can be useful in order to reduce nibbling of foundation by the bees by giving them only one frame of new foundation at a time.

Extractor: This is a mechanical device to spin the honey out of the combs by centrifugal force. It may be electrically powered or hand operated. Tangential extractors exert more force on the combs; care must be taken to avoid damage to combs. Radial extractors are gentler but will take longer and may be slightly less effective

Foam Closures: These are strips of foam plastic an inch or so thick and as long as the whole length of the entrance and a little more. They are now almost universally used to close up the hive for travelling, or if there is a toxic spraying alert and you need to keep your bees in. Screens must be in place above the top box in place of the crown board and roof in order to provide ventilation. The roof can be placed askew if rain is expected.

Foundation: Thin sheets of beeswax imprinted with hexagons ready to be built out into cells by the bees. Also available with drone hexagons instead of the usual worker size. May be normal thickness, or, less chewy, thin foundation. Supplied for brood (deep) or super (shallow) sizes, and often wired for extra strength for brood and extracted honey shallow frames.

Hiving-Off: This probably refers to splitting a colony in order to set up another hive. See chapter three on swarm control.

Hive straps: These are used to hold the boxes together, and except for the floor where lock slides are still used, straps are superior. They can be bought from d.i.y. stores, but the beekeeping suppliers' model can be cheaper and more suitable. They generally have a ratchet device with a quick release spring mechanism. Watch your fingers! Straps are used only when travelling with the hives. The weight of the boxes, with perhaps a brick on the roof top is usually sufficient to keep the hive together on its site.

Honey Flow: This is when the bees are collecting a lot of nectar from a large source. Bees require lots of productive flowers to yield significant amounts of honey, for example, a field full of the indigenous small white clover or several mature sycamore trees in flower.

Mouseguards: These are strips of perforated zinc, old queen excluder strips or other material which are placed right along the hive entrance to prevent mice getting into the hive and destroying it during the winter. Small slips of wood ¼" thick are placed into the entrance to give the correct bee-space above the floor for the mouseguard. The mouseguard is then pinned into position, and the slips of wood can then be removed. The gap is then just wide enough for the bees, but too narrow for the mice.

Nasenoff organ: This is a gland situated in the tail of the bee which can emit a pheromone which signals the position of the hive or swarm to bees which are airborne.

Out apiary: An apiary or site more than two miles from your main apiary in order to minimise the tendency of bees to fly back to their original site.

Parent colony or stock: This is the hive from which the queen has been removed, usually to make some form of artificial swarm.

Shake of Bees: The frame covered with bees is firmly shaken over the brood box or nucleus box in order to remove all the bees to their new home. One or more firm shakes downwards is the appropriate technique or the back of the hand holding the frame by a corner can be given a firm thump with the other hand which is useful to remove all nurse bees.

Shallow: Refers to super or honey crate also to size of frames and foundation.

Skep: This is sometimes referred to as a 'riskie' meaning basket. It is generally a dome shaped basket of closely twisted straw which is used to catch a swarm and provide temporary accommodation. Large tall skeps used to be the traditional hives, but are now superseded by the modern movable frame hive which allows for the easy removal of honey, and the effective management of the bees.

Snelgrove board: This is generally a crown board with three of the edge strips cut through twice at an angle so as to make three hinged gateways for the bees. These gateways can then be opened or closed by the beekeeper in order to drift bees successively to the swarm stock below over several days.

Starter strips: Narrow strips of thin foundation are attached to the top bars of frames instead of whole sheets. This usually helps the bees to draw natural comb straight, and aids in 'cut comb' production.

Super: A honey crate placed above the brood box, usually filled with shallow frames and foundation.

Swarm: This is a natural procedure in the hive when the colony is ready to reproduce the species by founding a new colony. The 'top swarm' is headed by the queen of the parent colony and that queen is likely to be a year or more old. A 'cast swarm' is usually smaller and is headed by a virgin queen. Top swarms generally issue forth nine days after the egg was laid in a queen cell cup. A cast swarm will follow seven days later or soon after, the virgin queen hatching on the sixteenth day since the egg was laid.

Swarm Stock: When you make an artificial swarm, the colony with the old queen and shaken bees in it is called the swarm stock to distinguish it from the colony with the brood and filled combs, the original hive or parent stock.

Uncapping Knife: Bees usually seal off the top of a cell full of honey with a wax capping. To extract the honey from the cells, these wax cappings are cut off with the uncapping knife so that the honey can run out in the extractor.

Index.

Acarine	58	Granulation	45
American Foul Brood	56	Heddon'	28
Apistan	48	hive	8
artificial shaken swarm	27, 28	hive tool	14
artificial swarm	28	Hoffman	12
Assembling your hive	12	Honey Harvest	41
autumn feeding	53	insecticides	58
Bayvarol	48, 54	inspection of hives	22
beeswax	52	June Gap	47
brace comb	12	Langstroth	10
Braula	59	laying workers	48
brood chamber'	5	location	48
cast	20	manipulating cloths	6
cast'	24	Manley	12
Chalkbrood	57	matchbox introduction	31, 49
Clip and mark queen	20	mating	47
Cottage hive	10	mouseguards	54
crown board	6	Nasenoff	15, 16
cut-comb honey	7	National	10
Cut-comb production	37	Nosema	57
cutting box	37	nucleus box	26
Dadant	10	nucleus hive	16, 18
Demaree	30	oil seed rape	7, 43
drifting	28	out apiary'	28
Drone cells	35	oxalic acid	32, 55
dummy boards	27	Pagden	28
dysentery	47	Paralysis	57
eke	13	Pesticides	58
emergency queen cells	28	Porter escape	42
equipment	9	propolis	43
European Foul Brood	56	queen cell cups	23
extracted honey	7	queen cells	23, 24
extracting	44	Queen cells	37
feeding	53	queen cups	37
Formic acid	32	queen excluder	19
foundation	5	Queen introduction	48, 49
frames	10	Queen Rearing	27
Frames and Bee-space	12	queenlessness	48
Glen	10	regulations	46
gloves	6	robbing	30

screen boards	51	swarm	15
sections	8	Swarm Control	24
Selling Honey	46	swarming	15
shake	27	Top Bar hive	8
Shaking bees	27	Travelling with Bees	50
shallow'	5	uncapping knife	44
site	6	uniting	49
Smith	10	uniting of colonies	8
smoker	7, 14	varroa control	32, 47
Snelgrove	29	varroa treatment	55
Snelgrove board	29	W.B.C.	10
spring syrup	47, 55	Worker cells	20, 39
starter strips	7	Manipulating cloths	23
super	5	'dud' queen cells	48
Supering	19	'emergency queen cells'	25

Frequently asked questions.

Some of the instruction given here also occurs in the chapters of the booklet. These F.A.Qs. can serve as reminders and may present information in a more immediately accessible form than in the main text.

1. I'm a complete beginner, how can I learn the craft of beekeeping?

You should join a local beekeepers' association where you will usually find there are summer demonstrations organised, and mentors may also be available for which there is often no charge. Local associations also often organise courses which may prepare you for certificate examinations. Many associations have winter programmes of lectures. Some agricultural colleges run courses in beekeeping with some courses leading to the National Diploma in Beekeeping, regarded as the high professional qualification.

2. I want to start beekeeping immediately. Where can I get my bees?

The best bees are probably local bees, accustomed to your particular conditions in your area. Your local beekeepers' association will often provide nucleus stocks of bees, or lists of people who want to sell nucleus stocks at reasonable prices. If you know how to catch a swarm, you can register with the police and with your association, and begin with a swarm which should be treated immediately you get it against the parasitic varroa mite. You can also buy a stock of bees from the beekeepers' supply companies, but the price is usually high, and the bees may not be ideally suited for local conditions.

3. The hive shows signs of swarming preparations, but I can't find the queen in order to carry out swarm control manoeuvres. What shall I do?

You could divide the colony and set up another hive. The best way is always to follow the natural way of the bees as far as possible, yet consistent with the aims of the beekeeper. Naturally, a swarm has no brood, but has a queen either laying or a virgin queen. Shake practically all the bees off all the frames into your new hive which is then positioned on the site of the original hive. Assuming that your inspection is not too late, and the hive has already given off a 'top swarm', the queen will then be in the new hive somewhere. Remember not to shake the frame with your chosen remaining queen cell on it so that the developing queen is not shaken off her food supply, but inspect that frame very carefully to ensure that only one queen cell and no laying queen remains in the parent colony. Move the parent hive with all the brood to the side, with the entrance facing away from the original colony entrance, in other words, practice either the Pagden or Heddon method of swarm control. Remember to remove all queen cells except the chosen one in the parent colony six or seven days after this manoeuvre. By shaking all the bees off the frames, the parent colony might be deprived of sufficient nurse bees to cover the brood. Therefore it may be wise to ensure that the last strong shake of each frame is given over the parent hive and not into the new hive which, having no brood, does not require any nurse bees in the meantime. Nurse bees generally cling more strongly to the frames.

4. My bees have swarmed, or, a queen cell has hatched. I want to make a nucleus stock, but can't find the queen. What shall I do?

It is really much easier to make up your new colonies with queen cells instead of trying to find a virgin hatched queen. That requires you to go into the hive before the new queens have hatched. But this is a fairly common predicament after the prime or 'top' swarm has come out, and the cast swarm is due very soon. The situation is very similar even if an artificial swarm has already been made a week or so earlier. Virgin queens are typically frisky, smaller than laying queens and usually difficult to find especially in a populous hive. There are several options, but none certain of success as virgin queens can fail to get mated or even get lost. Option a: Assume that the hatched queen has run away from the light below the uncovered frames, and make up your nucleus colony in the normal way, inspecting each frame thoroughly to ensure that the virgin queen remains in the parent hive. Insert a frame with a queen cell into your nucleus stock. Option b: Make a second artificial swarm or cast shaking all the frames of bees into another hive. The virgin queen will be in there somewhere. This new colony can be placed on the parent's site, and the parent moved away (Heddon of Pagden method). Only one queen cell is left in the parent colony. The new artificial cast swarm colony is therefore strengthened with bees from the parent hive, but it may be best to wait a day or two before moving the parent hive to avoid depleting that hive too much after shaking all the bees off the frames. Your strategy is important: do you want a strong swarm stock, perhaps to build new comb, or do you want to maintain the strength of the parent colony with its stock of brood to get away for a late summer honey flow? Whether the parent colony is on or off the original site can be important in your plans.

5. I left a queen cell in the parent colony, or, I left queen cells or a cell in my nucleus colony but mating of the virgin queen appears to have failed. What should I do?
You should wait at least three weeks before expecting to see your new queen laying successfully. If the weather has been very wet and windy the queen may take as long as a month to get mated and start laying. If the worker cells are well 'polished', there is usually a queen present. If the cells look uncared-for you may anticipate that all is not well; the queen may have got lost. If a month has elapsed, it is usually fairly safe to assume that mating has been unsuccessful, and the best thing to do is to unite the colony with a queen right hive using the newspaper method of uniting. You can test for queenlessness by inserting a frame with new worker eggs from another hive into the doubtful hive. Remember to shake most of the bees off the frame before inserting it to avoid fighting. Nurse bees still clinging will help to raise the queen cells and will generally not fight. If queen cells are drawn out from emergency cells after four or five days, the colony is definitely queenless. It is probably best to remove all except one new queen cell before uniting in case the bees are still prone to swarming. If it is late in the season, removal of excess queen cells will not be necessary and you can safely leave the hive to re-queen itself from the emergency queen cells and the strongest queen will destroy the others. Of course, if a queen-right colony is available that is the quickest and surest way to save the queenless hive by 'newspaper uniting' and the emergency cells should be removed first to ensure that an emerging virgin queen does not kill your laying queen.

6. I'm not sure whether my hive has a queen. How can I tell?
The simplest way to be sure instantly is to inspect the frames carefully. If there are eggs in the cells, or even larvae, then you can assume that the hive is queen-right. At the end of the season the queen may be 'off the lay', in which case you may need to test as in paragraph 5 above.

7. When can I expect my bees to swarm?
There are two answers to this question: a) what time of year. b) which day could they be expected to issue from an untreated hive.

Bees may be expected to prepare for swarming as soon as there are plenty of bees in the hive with plenty of food coming in. In Scotland, the main honey flow is usually around the middle of May when the apple blossom is still blooming and the sycamores are yielding lots of nectar. With global warming, the swarming season is starting progressively earlier, and swarms may be expected from around May 10th onwards. In the old days swarms were typically issuing from the beginning of June onwards. If you study the development of the bees from the egg until hatching, you will remember that the cell can be expected to be sealed on the eighth day after the egg was laid: 'three days eggs, five days larvae, sealed on the eighth, hatching on the sixteenth for the queen, twenty second for the worker and twenty fifth for the drone'. The swarm can be expected to leave an untreated hive soon after the first queen cell is sealed i.e. eight days after the egg was laid in a queen cell cup. The author has, however, noticed that nowadays, if the weather be fine, the swarm may come out before the eighth day. Possibly the hive becomes 'impatient' or possibly a queen cell is sealed slightly before the anticipated day. It is therefore wise to make swarm control measures as soon as eggs are seen in cups.

8. A swarm has settled in a highly undesirable place, what can I do to collect them?
If you can reach the bees you should place your nucleus box or your skep nearby, and if the bees cannot be shaken into the skep or box, you should gently pick up a handful or two of bees and shake them at the entrance to your skep which is open side down on a large board with an even larger sheet underneath. It helps the bees to go in if you prop up the entrance with a stone or small piece of wood. The sheet should be capable of being knotted on top of the skep in order to close the skep and board for transportation to your hive. Immediately, you should persistently smoke the bees remaining in the unsuitable place until all the bees are airborne and leaving the place where they had settled. The bees in the skep will normally fan the pheromone to bring in all the airborne bees.

9. Bees have built a hive in a place where they cannot be managed. What can I do?
You probably want the bees, so you should try to drive them out of their present home and into your hive. To do that you will require a good supply of smoke for a fair amount of time. You should smoke them as normal, and then bang on the sides of their home persistently and give more smoke, if possible above the brood nest and drive them into your hive which is fitted with frames and foundation, or better still, combs. Eventually the bees will get the message, leave their home and settle down in your hive. An alternative method is to make it impossible for the bees to return to their home, but freedom to fly out. The entrance to your hive, which should have a queen in it, is placed just beside the entrance to the colony you are depleting with your one-way trap. When the errant colony is entirely depleted, you can remove the trap and the bees will then rob out their original honey stores.

10. What should I do if I get stung?
A bee sting is hypodermic so applications to the surface of the skin are of little or no use. The best procedure is to scrape off the sting as quickly as possible so that the poison sac will cease to pump poison below the skin and the sting itself will cease to irritate. Blow smoke onto the area just stung to mask the pheromone and avoid attracting more stings to the site. If you have a bad reaction to bee stings it may help a lot to have some anti-histamine drug ready at hand in order to reduce the body's excessive reaction to the sting. In an emergency, when there is anaphylactic shock and breathing becomes difficult, the remedy should be provided by a doctor or other qualified person because the remedy is sub-cutaneous adrenalin administered in exactly the right way and in exactly the right dose because error can also

be fatal. If you get a few stings fairly regularly the body generally tolerates these pretty well, and is probably prophylactic for some complaints. It may be remembered that in some countries bee stings are seen as therapeutic against arthritis. Bee venom is sometimes collected from special plates which administer minute electric shocks to the bees. The venom is then collected and made into a drug which is injected for therapy.

11. I have been offered hives of another type from the one I already possess. Should I take them?

It is best to keep to one type of hive because you will frequently need to move frames from one hive to another. If the hive types are mixed you will obviously encounter many problems. The author's preference is for the Smith hive which has a top bee space and takes up less space in the car or in storage than most other practical hive types.

12. Grandmother used to kill the queen when she found her hive was making queen cells and intending to swarm. Should I do the same?

No, it is a waste to kill a perfectly good queen. If you can't use her, put her in a matchbox with six workers and a tiny dab of honey and give or sell her to somebody else. It is best to have a nucleus hive for each hive that you have, as a back-up in case the virgin queen in the parent colony fails to get mated or gets lost. Instead of killing the queen, remove her and carry out an artificial swarm. The occasion for culling the queen is when a colony is always foul-tempered or is suffering from some particular disease and beekeeping becomes unmanageable. Usually a fractious colony becomes easily manipulated as soon as the queen is changed. Some diseases disappear when the queen is replaced.

13. When should I remove the queen for swarm prevention?

If the queen cells in the hive are at the larval stage the queen should be removed immediately to avoid the beekeeper being 'caught out' and losing a 'top swarm'. If the hive is only at the 'eggs in cups' stage it is probably best to wait another four days to see if the bees are really intending to swarm. However, if your main goal is increase of colonies, you can encourage the bees to go all out for new queens by doing an artificial swarm at the early 'eggs in cups' stage.

14. Should I cut out all the queen cells except one as soon as I remove the queen?

It depends what stage the queen cells have reached in their development. If the cells are still at 'eggs in cups' stage it is best to leave them for a few days until they have reached a day or two into the larval stage. A cast swarm with a virgin queen will not issue from the hive for at least sixteen days after the egg was laid, which will probably be nine days after a top or first swarm has gone out or has been conducted by the beekeeper. If the beekeeper cuts out queen cells too early, the bees will simply make more queen cells by converting worker larvae to queen larvae and creating the hard-to-find 'stubby' or 'emergency' queen cells. The wise beekeeper cuts the queen cells only when the worker larvae are too old to convert into emergency queen cells. Six or seven days after the queen was removed is a suitable time to remove the unwanted queen cells and therefore prevent cast swarms.

15. When should I feed my bees?

If you have just acquired your bees and they have to build comb, you should feed them on half sugar and half water in cool weather, or one part sugar to two parts water when flowers and pollen are available. The feed is made into warm syrup, not hot. If the weather is poor, or if there are few nectar producing flowers about, you should also feed your bees unless you know that they have two or more brood frames full of honey stores. The hive can also be 'hefted' to ensure that there are enough stores present. Bees are generally fed on 'winter syrup' in September, the aim being to have all the syrup converted by the bees before the cold weather sets in. Spring syrup is generally given from the middle of March onwards until adequate forage becomes available with plum, cherry currant and dandelions coming out. Emergency winter feeding is usually given in the form of sugar candy directly on the frame tops.

16. When should I treat my bees for infestation by the varroa mite?

If your bees are without brood, as they will be if you have a swarm or package bees, you should treat them immediately with 3.5% oxalic acid in 20% sugar solution, 30ml. trickled between the frames to soak the bees for each hive treated. The same treatment should be given when the bees are without brood, or almost so, twenty five days after the queen was removed in an artificial swarm. The same treatment should be given in the middle of winter to mop up as many surviving varroa as possible. Treatment is given on a mild day when the bees are flying, after gently smoking the bees in the usual way. At the end of the summer, it is usual to treat the hive with a commercial varroa killer, but with increasing resistance to these chemicals recourse may be required from other agents to control the mites. Until bees resistant to the mites become widely available, leaving the bees without treatment is probably not an option. It is advised that oxalic acid be only given at intervals of at least three months.

17. Should I filter my honey?

No, honey should not be filtered. To do so would remove the tiny pollen particles which are believed to be therapeutic and helpful to reduce pollen allergies. Honey should be put through a fine sieve or muslin when it is warm and runs freely.

18. I want to sell some honey but don't know what to charge for it.

You can get some idea of appropriate prices from shops and your local beekeepers' association. Sometimes other beekeepers are offering their honey far too cheaply. Remember all the hours of work you and your bees have put in, and remember that the honey from bees in a cold climate is always bound to be higher priced than imported honey from hot countries where yields are usually much higher. The government beekeeping web site usually has details from national surveys of honey prices, and there are other sites too which will appear if you put 'honey prices' into your search engine on the internet. If you sell to a retailer, you will have to allow for the retailer's profit margin.

19. Why are there so many drone cells in my hive?

It should be understood that drones come from an unfertilised egg, unlike queen and worker eggs which are fertilised. Sometimes the queen is either not mated properly, or she has simply run out of sperm in her spermatheca. The need is then for a new, fertile queen.

20. It is late in the season (late August and September in Scotland). My hive appears to be queenless, is it really queenless?

Commonly in the late part of the season a hive can appear to be queenless, with the queen going 'off the lay' because of bad weather or a dearth of nectar and pollen. Remember that bees need lots of gardens and many trees and fields full of blossom to get enough food. One garden is rarely sufficient for their needs. If your hive is weak you will need to give lots of 1:2 sugar-water syrup, and possibly artificial pollen-substitute to build up sufficient strength to see your bees through the winter. As soon as there is a little more strength in the hive, give 1:1 winter syrup before the end of September so that the hive has adequate stores. If cold weather has already set in, candy will need to be given because the bees will not have time to convert all that syrup. If there's still no sign of eggs or larvae two or three days after giving food, test for queenlessness with a frame of worker eggs as indicated in question no.5 above.

21. I have found a lot of dead bees in front of my hive. What has happened?

There are a number of possibilities. The least serious is that there has been an invasion of bees from another colony and there has been a bit of a battle between your bees and the invaders. That can be a good thing for the beekeeper and the hive if the invaders are a large force of bees of a good character. The dead bees spell trouble if there is a sudden outbreak of disease, but usually the bees die away from the hive. It is a bad sign if the bees are dead because of locally sprayed pesticides, and it is advisable if the cause is unknown to send a sample of bees to SASA in Scotland or to your local environmental health agency if the hives are situated elsewhere.